Translational Medicine and Bioengineering Conference 2016

San Francisco, California, USA
12 – 13 November 2016

ISBN: 978-1-5108-6374-3

Printed from e-media with permission by:

Curran Associates, Inc.
57 Morehouse Lane
Red Hook, NY 12571

Some format issues inherent in the e-media version may also appear in this print version.

Printed by Curran Associates, Inc. (2018)

For permission requests, please contact AIChE
at the address below.

AIChE
120 Wall Street, FL 23
New York, NY 10005-4020

Phone: (800) 242-4363
Fax: (203) 775-5177

www.aiche.org

Additional copies of this publication are available from:

Curran Associates, Inc.
57 Morehouse Lane
Red Hook, NY 12571 USA
Phone: 845-758-0400
Fax: 845-758-2633
Email: curran@proceedings.com
Web: www.proceedings.com

TABLE OF CONTENTS

BIOPHARMACEUTICALS

GENE AND DRUG DELIVERY

IMMUNOENGINEERING

BIOMANUFACTURING

STEM CELLS REGENERATIVE MEDICINE

POSTERS

Engineering Antibody Fabs for Long Acting Delivery to the Eye

Conference: Translational Medicine and Bioengineering Conference

Year: 2016

Proceeding: Translational Medicine and Bioengineering Conference

Group: General Submissions

Session: Biopharmaceuticals

Time: Saturday, November 12, 2016 - 4:40pm-4:55pm

Authors:

Kelley, R., Genentech, Inc.

Maximal clinical benefit for treatment of back of the eye disorders such as agerelated macular degeneration (AMD) with protein therapeutics is currently obtained with intravitreal injections at 4-8 week intervals. Use of long acting delivery (LAD) technologies or molecules with increased vitreal half-life would benefit patients through reduced dosing frequency, less incidence of injection-related complications, and perhaps improved durability of treatment outcome. Challenges for LAD to the eye or clearance modification include restrictions on volume (â¤100 Î¼L) of injection necessitating high concentration formulations, a short list of formulation excipients approved for ocular use, stability of proteins to vitreous humor conditions, and a limited understanding of the mechanism of protein clearance from the eye. We have been using pharmacokinetic (PK) studies in a rabbit model to explore molecular attributes of antibody Fab fragments affecting vitreal half-life. We find that hydrodynamic size is a key contributor to the rate of vitreal clearance, suggesting that diffusion out of the vitreous cavity is a rate-determining step, with vitreal halflife showing a linear dependence on hydrodynamic radius calculated from light scattering measurements. Increasing the hydrodynamic size of a protein therapeutic slows clearance and is an effective means to achieve less frequent dosing. In addition, we present a case study involving the re-engineering of a therapeutic Fab to increase the stability to vitreous humor conditions yielding variants that are more suitable for LAD.

Evaluation of Invertebrate Hemoglobins As Novel Blood Substitutes

Conference: Translational Medicine and Bioengineering Conference

Year: 2016

Proceeding: Translational Medicine and Bioengineering Conference

Group: General Submissions

Session: Biopharmaceuticals

Time: Saturday, November 12, 2016 - 4:25pm-4:40pm

Authors:

Elmer, J., Villanova University
Zimmerman, D. C., Villanova University
Dienes, J. A., Villanova University
Neely, C., Villanova University
Nicholas, M., Villanova University
Zabarnick, S., Villanova University

While donated blood is the preferred treatment for hemorrhagic shock, it has a relatively short shelf life (42 days) and it must be constantly refrigerated at 4°C. Consequently, donated blood is often unavailable in remote areas where it is needed most (e.g. military battlefields). One way to address this problem is by developing âblood substitutesâ that mimic the most important functions of blood (e.g. O_2 transport) but are much more stable than red blood cells. Most previous blood substitutes have been based on human or bovine hemoglobin, since hemoglobin is an excellent O_2 transporter and it can easily be modified to increase its stability. Unfortunately, while these hemoglobin-based oxygen carriers (HBOCs) showed early promise, they all eventually failed clinical trials due to an increased risk of severe adverse events including hypertension, stroke, and heart attack. We suggest that all of these problems can be attributed to removing the naturally intracellular human hemoglobin from the protective environment of the red blood cell and exposing it to the harsh conditions in the bloodstream. Therefore, our hypothesis is that the naturally extracellular hemoglobins (also known as erythrocruorins or Ecâs) found in some invertebrates (e.g. earthworms) may be much better blood substitutes. In this study, we have purified and compared Ecâs from a variety of organisms, including worms (e.g. Lumbricus terrestris), leeches (e.g. Nephelopsis obscura), clams (e.g. Astarte castanea), and a snail (Biomphalaria glabrata). Our results show that these erythrocruorin-based O_2 carriers (EBOCs) are highly stable (T_m = 50-60°C), resistant to oxidation during storage and in vivo, and possess a wide range of O_2 transport properties (e.g. P_{50} = 2-20 mm Hg at 25°C). In addition, we have also covalently modified earthworm hemoglobin in two ways: (1) glutaraldehyde cross-linking and (2) conjugation to polyacrylic acid. The cross-linked hemoglobin is highly stable (T_m = 70°C), while the PAA-conjugated hemoglobin is actually able to resist denaturation in the autoclave at 121°C for 30 minutes. Finally, preliminary animal studies by our group and others also show that Ecs can be transfused into mice, rats, and hamsters without any of the harmful side effects observed with previous HBOCs. Overall, these results suggest that EBOCs may be safe and effective blood substitutes that are stable enough to be deployed in areas where donated blood is unavailable.

Translatable Affibody Ligand Discovery: Engineered Biopanning, Deep Sequence-Guided Library Design, and Pharmacokinetic Modulation

Conference: Translational Medicine and Bioengineering Conference

Year: 2016

Proceeding: Translational Medicine and Bioengineering Conference

Group: General Submissions

Session: Biopharmaceuticals

Time: Saturday, November 12, 2016 - 3:55pm-4:10pm

Authors:

Hackel, B. J., University of Minnesota
Stern, L. A., University of Minnesota
Woldring, D., University of Minnesota
Case, B., University of Minnesota

The affibody is a 58-amino acid three-helix bundle that has been an effective ligand scaffold for clinical molecular imaging probes, targeted therapy, and biotechnology. A key challenge in discovery and development of affibodies, and all protein ligands, is the inconsistency of binders evolved against recombinant target to translate to strong, specific binding to genuine intact cellular target. Direct selection of yeast-displayed antibody libraries on adherent mammalian cell monolayers has been previously used to address this limitation. To enhance the robustness and selective control of cellular panning, we have systematically elucidated the elements that dictate ligand enrichment. Introduction of an extended flexible PAS peptide between the cell-surface tethering protein and the displayed ligand enhances enrichment 44-fold, which enhances discovery of all binders in a library. Moreover, use of mammalian cells with moderate target expression enables modest affinity discrimination to selectively enrich the strongest ligands. We have demonstrated controlled chemical cleavage of displayed ligands to control valency over two orders of magnitude, which empowers strict and broadly titratable affinity discrimination. Twenty different ligand discovery campaigns were competed, in two scaffolds (affibody and fibronectin domain) and multiple targets, to compare various selection strategies. Direct cell panning enhances discovery of translatable hits four-fold versus campaigns using initial enrichment on magnetic beads coated in recombinant target followed by either flow cytometry with recombinant target or biotinylated cell lysate. Conditions to optimize depletion panning, to remove undesirable binders, will be discussed.

Deep sequencing an epitope-diverse pool (>37,000 unique variants in >5,000 family clusters) of evolved binders identified amino acid motifs consistent with evolutionary efficacy, which revealed a non-spatial gradient of diversity (Shannon entropy: 1.4-4.0) in an optimized paratope library. Implementation of these sitewise amino acid distributions yielded a 50-fold increase in the binder discovery efficiency towards a broad panel of targets. Moreover, affibodies from this constrained library design exhibit dramatically higher thermal stability (24Â ±12Â°C higher T_m) than binders from a naÃ¯ve library.

In parallel, the affibody framework was engineered for reduced charge density while retaining stability, solubility, and binding. A synthetic consensus design strategy â merging directed evolution and bioinformatics of evolutionarily tolerant residues â yielded better variants than directed evolution or informatics alone. Neutralization of three acidic and three basic residues resulted in a 39Â±10% reduction in non-specific renal retention while increasing tumor targeting 21Â±5% in murine xenograft models.

Collectively, enhancements in library design, ligand selection, and framework design for biodistribution improve the ability to engineer translatable affibody ligands.

Biopharmaceutical Informatics: A Strategic Vision for Improved Translation of Biologic Drug Discoveries into Drug Products

Conference: Translational Medicine and Bioengineering Conference

Year: 2016

Proceeding: Translational Medicine and Bioengineering Conference

Group: General Submissions

Session: Biopharmaceuticals

Time: Saturday, November 12, 2016 - 3:40pm-3:55pm

Authors:

Kumar, S.
Singh, S., Pfizer

Research scientists discover several promising novel drug candidates every year, but only a few of these newly discovered compounds reach the clinic after several years. The current rate of translating drug discoveries into medicines is approximately 11%. A major portion of the developmental costs of new drugs is incurred during clinical trials, whose outcomes depend on the choices made during the selection (design) of the drug molecule, formulation, manufacturing, dosing, clinical trial design and patient population selection. The development projects that yield drug products, successfully, must recover costs for their development and those of the discontinued ones. This leads to high prices for the innovative drug products. The low translation rates and high development costs make drug discovery and commercialization a risky affair and point to inefficiencies in drug discovery, development and testing. While biological activity is justifiably the focus during the discovery of biopharmaceutical candidates, the macromolecular sequenceâstructural properties of these candidates can also inform us about challenges in biologic drug development. These include cell line expression levels, purification column behavior, potential physicochemical degradation routes such as aggregation and chemical degradation hotspots, interactions with extractables and leachables, behaviors of highly concentrated solutions, immunogeÂnicity, pharmacokinetics/dynamics, and so on.Â Such insights, when they come via computation at early stages of lead candidate design or selection, can significantly improve efficiency of biologics drug disÂcovery and development.
In this talk, we intend to present our strategic vision for Biopharmaceutical Informatics. Biopharmaceutical Informatics combines information technology (database creation, curation

and analytics) with computational biophysics (molecular modeling, simulations and structural bioinformatics) to help understand challenges faced during drug product development. For example, Biopharmaceutical Informatics can help selection of biologic drug candidates with favorable biophysical attributes at the early stages of product development. We have been building in silico tools that can rapidly identify biologic drug candidates with product development issues at early stages and suggest sequence optimizations for lead candidate(s) in a project. Early implementation of such predictive tools can shorten project timelines, save resources and avoid lengthy detours when unanticipated challenges are encountered later during product development. Potential applications of Biopharmaceutical Informatics at the later stages include improvements in cell-line productivity, drug substance purification yields, design of novel formulation excipients and assessing the impact of sequence variations on drug product quality.

Engineered Multivalency for Enhanced Efficacy of HER3-Targeted Affibodies

Conference: Translational Medicine and Bioengineering Conference

Year: 2016

Proceeding: Translational Medicine and Bioengineering Conference

Group: General Submissions

Session: Biopharmaceuticals

Time: Saturday, November 12, 2016 - 4:10pm-4:25pm

Authors:

Jay, S.
Schardt, J.

HER3 (also known as ErbB3), a member of the HER receptor tyrosine kinase (RTK) family, is a potent mediator of tumor progression in breast, lung, ovarian, and other cancers. Additionally, HER3 has been implicated in mediating resistance to EGFR-, HER2-, and PI3K-targeted drugs (among others). Yet, while some candidate molecules have entered clinical trials, there are currently no FDA-approved HER3-targeted therapeutics. Even if one or several drugs targeting HER3 do eventually receive approval, there is still likely to be a large proportion of refractory patients; for example, currently approved HER2-targeted therapeutics typically show efficacy in only ~24%-64% of patients with HER2+ metastatic breast cancer in clinical trials. Thus, development of new HER3-targeted therapeutics would fill a critical clinical need in cancer therapy.

HER3 is unique amongst the HER family in that it has an inactive or very weak kinase domain, necessitating its partnership with an active RTK via dimerization to induce downstream signaling. In addition, HER3 exists in homo-oligomeric clusters on cancer cell membranes, with binding of the HER3 ligand Neuregulin 1B (NRG) leading predominantly to formation of potent pro-neoplastic HER3-HER2 heterodimers. Based on this knowledge, we hypothesized that multivalent ligands with multiple high affinity HER3 binding domains separated by a flexible peptide spacer could sequester HER3 into homotypic configurations that prevent its interaction with HER2 or other potential partners (e.g. EGFR, c-MET, IGF-1R), thus inhibiting pro-neoplastic signaling. To test this hypothesis, we first engineered a novel class of bivalent HER ligands and demonstrated that bivalent NRG can inhibit HER3 signaling in cancer cells from diverse tissue origins, inducing apoptosis and reducing chemokinesis. We have now developed a second generation of multivalent HER3 ligands based on engineered affibody domains that have increased translational potential compared to bivalent NRG due to reduced risk of pro-neoplastic side effects.

Crucially, multivalent HER3 affibodies are more effective at inhibiting HER3 signaling (HER3 and Akt phosphorylation) and reducing cell survival across multiple cancer cell types than monovalent versions of the same affibody. Mechanistically, multivalency endows HER3 affibodies with the ability to downregulate or degrade HER3, whereas minimal degradation is observed with monovalent affibodies. Current investigations are assessing whether this multivalent effect is specific to HER3 or can be applied to other RTK targets as well. Multivalent HER3 affibodies are well tolerated by mice and work most effectively in combination with other drugs, specifically trastuzumab and small molecule chemotherapeutics such as buparlisib. In summary, using an approach fundamentally different than the HER3-targeted therapies that are currently in clinical trials, we have established that HER3 sequestration via engineered multivalent ligands is a versatile approach with broad therapeutic potential in cancer.

Directed Evolution of New Viruses for Therapeutic Gene Delivery

Conference: Translational Medicine and Bioengineering Conference

Year: 2016

Proceeding: Translational Medicine and Bioengineering Conference

Group: General Submissions

Session: Gene and Drug Delivery

Time: Saturday, November 12, 2016 - 9:15am-9:45am

Authors:

Schaffer, D. V., University of California, Berkeley

Strong basic and translational efforts in the gene therapy field have culminated in successes in an increasing number of clinical trials involving viral vectors, particularly ones based on adeno-associated virus (AAV). These include trials for monogenic diseases such as hemophilia B, Leberâs congenital amaurosis, Sanfilippo B, and lipoprotein lipase deficiency (for which a gene therapy product was approved in the EU in 2012). AAV is thus capable of safe and therapeutic gene delivery to some cell and tissue targets; however, vectors in general face a number of challenges that limit their efficacy. These include anti-vector neutralizing antibodies, low transduction of some therapeutically relevant cells in vitro or in vivo, difficulty in overcoming cellular and physical barriers within complex tissue structures, and an incapacity for targeted delivery to specific cells. These challenges are not surprising, as nature did not evolve viruses for our convenience to use as human therapeutics, and thus âoff the shelfâ natural viruses do not meet a range of clinical needs.Â In most situations there is insufficient mechanistic knowledge of underlying virus structure-function relationships to empower rational design to improve such vectors; however, directed evolution has been emerging as a strategy to engineer novel viral variants that meet specific biomedical needs. We were the first to develop and have since been implementing directed vector evolution â the iterative genetic diversification of a viral genome and functional selection for desired properties â to address a number of problems with AAV. Genetic diversification has included the random diversification of peptide sequences at defined locations in the capsid, random point mutagenesis of the cap gene, and recombination of capgenes from a number of parental serotypes to create random chimeras. Using a range of in vitro and in vivo selection strategies, we have evolved AAV for evasion of neutralizing antibodies, enhanced biodistribution and spread within a target tissue, greatly improved delivery efficiency, and targeted delivery in vitro and in vivo, and the resulting vectors are being clinically translated to meet human therapeutic needs.

Discovery of Polymers for Nucleic Acid Delivery Using Combinatorial and Cheminformatics Methods

Conference: Translational Medicine and Bioengineering Conference

Year: 2016

Proceeding: Translational Medicine and Bioengineering Conference

Group: General Submissions

Session: Gene and Drug Delivery

Time: Saturday, November 12, 2016 - 9:45am-10:00am

Authors:

Rege, K., Arizona State University

The delivery of nucleic acids, including plasmid DNA, to mammalian cells has diverse applications in disease therapies, regenerative medicine, synthetic biology and protein production. Molecular carriers of plasmid DNA have met with limited success compared to viral vectors and some other non-viral vehicles. Despite availability of sophisticated delivery vehicles and investigations into biological factors, a comprehensive approach to the discovery of effective DNA carriers has likely hampered success in this field. Here, we propose a holistic approach in which we integrate exploration of diverse chemical space using library syntheses, cheminformatics-driven molecular engineering, and the discovery and abolition of intracellular resistances, in order to develop a novel discovery paradigm for polymer-nucleic acid delivery. Libraries of polymers and their derivatives were generated using aminoglycoside antibiotics as monomers. Support Vector Regression (SVR)-based Quantitative Structure-Activity Relationship Models (QSAR) modeling approaches were employed in order to elucidate the role of physicochemical properties on transgene expression efficacy of polymers. Given their role in cellular processes, we investigated the role of proteins of the kinome and epigenome in determining the efficacy of polymer-mediated transgene expression. Effective polymers were employed for the ablation of tumors using plasmid delivery, transgene expression in stem cells, gene silencing, and adenoviral transduction of resistant tumors.

Histone-Targeted Gene Delivery Carriers for Bone Regeneration

Conference: Translational Medicine and Bioengineering Conference

Year: 2016

Proceeding: Translational Medicine and Bioengineering Conference

Group: General Submissions

Session: Gene and Drug Delivery

Time: Saturday, November 12, 2016 - 10:00am-10:15am

Authors:

Munsell, E.
Sullivan, M., University of Delaware

Gene therapy has emerged as a promising treatment option for the regeneration of damaged skeletal tissue. Gene transfer applications offer non-invasive, low cost alternatives to conventional bone grafting procedures, with a capacity for sustained, controlled expression of nascent proteins that can initiate bone repair. However, the design of synthetic delivery systems that negotiate efficient intracellular trafficking and controllably bind and release DNA within the nucleus represents a significant challenge. Overcoming these hurdles requires a combination of well-controlled materials approaches with techniques to understand and direct cellular delivery. Recent investigations have highlighted the roles histone tail sequences play in directing nuclear delivery, nuclear retention, and DNA transcription in chromatin. In this work, we establish the ability to recapitulate these natural histone tail activities within non-viral gene carriers to improve gene transfer and expression of exogenous factors. We demonstrate that histone-targeted carriers employ vesicular retrograde trafficking pathways to reach the nucleus without endosomal escape, and we identify their key interactions with native histone effectors that initiate nuclear targeting and enhance gene expression. We also show that histone-targeting leads to improved osteogenic growth factor expression in mesenchymal stem cells, ultimately enhancing differentiation along cellular lineages essential to endochondral ossification and fracture healing. Most notably, we demonstrate that histone-targeted growth factor expression can trigger similar levels of differentiation as topically applied growth factors, yet requires only 1/100th the quantity of growth factor. Future work seeks to further enhance growth factor delivery and expression capabilities through the development of a tunable delivery scaffold designed to mimic additional aspects of the natural histone octamer architecture.

Multi-Stage Drug Delivery System for Enhanced Payload Delivery to Lymph Node Cells

Conference: Translational Medicine and Bioengineering Conference

Year: 2016

Proceeding: Translational Medicine and Bioengineering Conference

Group: General Submissions

Session: Gene and Drug Delivery

Time: Saturday, November 12, 2016 - 10:45am-11:00am

Authors:

Thomas, S. N., Georgia Institute of Technology
Schudel, A., Georgia Institute of Technology
Higginson, C., Georgia Institute of Technology
Finn, M. G., Georgia Institute of Technology

Targeted drug delivery to lymph nodes (LN), tissues where organized lymphocyte accumulation in the body occurs, has the potential to improve the treatment of a variety of pathologies, such as B and T cell malignancies and immunodeficiency viral infections, as well as enhance the efficacy of vaccines and cancer immunotherapy. To date, size-based principles of lymphatic-mediated transport have emerged as an effective means to achieve facile LN delivery of formulated agents after injection in the periphery. However, within the LN, solutes and particulates are filtered based on size, with smaller molecules penetrating more deeply into LN follicles and the paracortex than large particulates by virtue of their increased diffusivity and uptake within LN conduits. Herein, we report the design and implementation of a multistage lymphatic delivery and payload release system combining 30 nm thiolated poly(propylene sulfide) nanoparticles that exhibit prodigious uptake into lymphatics after administration with the thiol-reactive oxanorbornadiene (OND) linkage chemistry that degrades in a pH- and polarity-insensitive manner at programmable half-lives ranging from hours to days to release small-molecule cargo. Post administration in the skin, nanoparticle tethering dramatically enhances model cargo accumulation within LN draining the site of injection that when released via the OND penetrates deeply throughout the collagenous LN interstitium due to its increased diffusivity relative to polymer nanoparticles. As a result, small molecule cargo is delivered to cell subtypes localized within the LN cortex and paracortex at high levels, a 10- to 1000-fold relative to free cargo alone or cargo tethered to the nanoparticles via an irreversible linkage.

These data suggest that by utilizing an engineered multi-stage system to overcome barriers to both lymphatic and lymph node uptake, payload can be delivered to previously undruggable cell subpopulations resident within lymph nodes and has the potential to dramatically enhance drug bioactivity and resulting therapeutic effect.

Probing Cellular-Particle Interactions and Deformability in the Design of Vascular-Targeted Carriers

Conference: Translational Medicine and Bioengineering Conference

Year: 2016

Proceeding: Translational Medicine and Bioengineering Conference

Group: General Submissions

Session: Gene and Drug Delivery

Time: Saturday, November 12, 2016 - 10:15am-10:30am

Authors:

Eniola-Adefeso, L., University of Michigan

Localized delivery of therapeutics offersÂ theÂ possibility of increased drug effectiveness while minimizing side effects often associated with systemic drug administration. Factors that influenceÂ the possibility of targeted therapeutics to reachÂ the vascular wall includeÂ theÂ ability to 1) identify disease-specific target epitope expressed by theÂ vascular cells; 2) identify optimum drug carrier type, shape and size for efficient interaction withÂ theÂ vascular wall; and 3) identify a drug-carrier combination that allow for theÂ effective release of therapeutics atÂ theÂ intended site. Existing literature has focused on identifying target epitopes andÂ theÂ degradation/drug release characteristics of a wide range of drug-carrier formulations.Â Our work focuses on probing the role of particle deformability, material chemistry and blood rheology/dynamics onÂ theÂ ability of vascular-targeted drug carriers to interact withÂ theÂ blood vessel wall â an important consideration that will controlÂ theÂ effectiveness of drug targeting regardless of the targeted disease or delivered therapeutic. This presentation will highlight key carrier-blood cell and carrier-plasma protein interactions that affect drug carrier binding to the vascular wall and suggest the optimum drug carrier modulus and material chemistry for vascular-targeted drug delivery applications in a solid tumor microvascular network and in large vessels associated with many cardiovascular diseases.

Design of Polyelectrolyte Multilayers to Promote Immune Tolerance

Conference: Translational Medicine and Bioengineering Conference

Year: 2016

Proceeding: Translational Medicine and Bioengineering Conference

Group: General Submissions

Session: Gene and Drug Delivery

Time: Saturday, November 12, 2016 - 10:30am-10:45am

Authors:

Jewell, C. M., University of Maryland

In multiple sclerosis (MS), myelin insulating neurons is attacked by the immune system. New studies in human MS patients reveal excess inflammation through toll like receptors (TLR) contribute to disease, and further, that co-administration of self-antigens and regulatory signals can promote more specific tolerance. Biomaterials can enable this co-delivery and other attractive features (e.g., tunable loading, cargo protection), but can cause inflammation that could exacerbate autoimmunity. Thus, we designed immune polyelectrolyte multilayers (iPEMs) mimicking attractive features of biomaterial, but self-assembled entirely from myelin-derived peptides and a macromolecular, regulatory TLR9 antagonist. We hypothesized these iPEMs would reduce TLR-driven inflammatory cues that expand self-reactive T cells, promoting regulatory T cells that control disease without broad immunosuppression. We show iPEMs co-deliver each signal to dendritic cells, attenuate TLR9 signaling, deactivate DCs, and polarize antigen-specific T cells toward regulatory T cell function. Importantly, these effects depend on integration of both myelin and the TLR antagonist. In mice, iPEMs reduce inflammatory T_H17 and T_H1 cells, expand regulatory T cells, and eliminate MS-like disease. In samples from human MS patient, iPEMs also activate T cells, but polarize the cytokine profiles of these cells away from inflammation and toward tolerogenic function. This is the first time polyelectrolyte multilayers have been used to promote tolerance and could represent a new platform for combatting autoimmune disease by programming the combinations and relative combinations of self-antigens and regulatory TLR ligands.

Controlling Immune Cell Dynamics Using Microfluidics

Conference: Translational Medicine and Bioengineering Conference

Year: 2016

Proceeding: Translational Medicine and Bioengineering Conference

Group: General Submissions

Session: Immunoengineering

Time: Saturday, November 12, 2016 - 2:00pm-2:15pm

Authors:

Tay, S., UNIVERSITY OF CHICAGO

Immune cells receive time-varying signaling inputs, use gene regulatory pathways to process these signals, and generate outputs by secreting signaling molecules. Characterizing this input-output dynamics helps understanding the underlying regulatory mechanisms and allows manipulating immune interactions for a desired therepeutical outcome. Significant variability in molecular parameters between cells makes time-dependent single-cell and single-molecule analysis crucial in understanding how biological systems operate. I will talk about how we developed automated, high-throughput microfluidic single-cell and single-molecule analysis systems with unprecedented capabilities and measurement accuracy, and how we use them in understanding immune coordination during response to infection. Our recent efforts have resulted in a new set of technologies, including microfluidic systems to measure single-cell cytokine secretion dynamics, cell culture systems that create programmable diffusion-based chemical gradients, devices to measure cell-cell communication via secreted factors, and a new method for digital quantification of proteins and mRNA in the same cell. I will also talk about new biological insight emanated from our experimental and modeling efforts on how single-cells detect and encode dose and frequency information of an input signal using the immune pathway NF-Î°B, and how molecular noise improves dynamic signal encoding and decoding, and how gene expression and cell fate can be controlled by modulating the temporal profile of biochemical input signals. The primary goal in our combined technology/cell biology effort is to help develop computer models of tissue-level immune response that will serve as a test-bed for drug or genetic perturbations on disease pathways.

Dendritic Cell-Targeted Immunomodulation for Tolerance

Conference: Translational Medicine and Bioengineering Conference

Year: 2016

Proceeding: Translational Medicine and Bioengineering Conference

Group: General Submissions

Session: Immunoengineering

Time: Saturday, November 12, 2016 - 1:30pm-1:45pm

Authors:

Keselowsky, B., University of Florida

This presentation will highlight results from our biomaterials and molecular based approaches to direct the immune system toward tolerance. Microparticle-based systems have been designed as injectable vaccines to retrain the immune system to correct aberrant activation toward self-antigens. Hydrogels and microparticles with encapsulated immunomodulatory factors and antigen provide targeted, controlled delivery in vivo to both intracellular and cell surface targets of immune cells, dendritic cells in particular, in order to promote tolerance. Using a microparticle formulation consisting of a combination of suppressive factors we have demonstrated prevention and reversal of type 1 diabetes in non-obese diabetic mice, as well as treatment of multiple sclerosis in an EAE mouse model. Investigations into the immunological mechanisms implicate a role for regulatory T cells and B cells. In addition, we are developing molecular engineering approaches to direct tissue-localized metabolism to guide immune cell function toward tolerogenic phenotypes.

Immunoengineered Particles As Artificial Immune Cells: the Importance of Size and Shape

Conference: Translational Medicine and Bioengineering Conference

Year: 2016

Proceeding: Translational Medicine and Bioengineering Conference

Group: General Submissions

Session: Immunoengineering

Time: Saturday, November 12, 2016 - 1:45pm-2:00pm

Authors:

Green, J., Johns Hopkins University

Biodegradable materials, including polymeric particles, can be engineered to mimic attributes of biological cells. Our lab is especially interested in the design and development of immunoengineered artificial antigen presenting cells (aAPCs). We construct biodegradable poly(lactic-co-glycolic acid) (PLGA) micro- and nanoparticles of various sizes and then alter their geometry through the use of an automated multidimensional film-stretching device. We functionalize the surface of these particles by conjugation of a biomimetic Signal 1 protein, an MHC Db IgG gp100 dimer that gives antigen specificity, and a Signal 2 protein, an anti-CD28 antibody as an activation signal. We have observed that the interaction of a T Cell with the surface of an aAPC can be dependent on the physical properties of the particle including its size and geometry. We find that with the same protein coverage, anisotropic particles as aAPCs are more potent than spherical aAPCs. In particular, oblate ellipsoids (disc shaped), with a large radius of curvature, lead to more effective aAPC activity than prolate ellipsoids (football shaped), which in turn are more effective than spherical particles. Size, another physical property, plays a similar role with microsphere-based aAPCs more effective than nanosphere-based aAPCs. Non-spherical, anisotropic shapes were particularly important for nanoscale aAPCs, and anisotropic shapes increased cytotoxic T Cell proliferation in vitro and in vivo, reduced non-specific cellular uptake in vitro, and led to longer blood circulation half-life in vivo. We find that microparticles synthesized utilizing a flow-focusing microfluidic device can lead to spherical and anisotropic aAPCs with a polydispersity index of 1.05, indicating a nearly uniform population. We also find that this technology can be extended to fabricate anisotropic particles coated with supported lipid bilayers. These hybrid particles can allow for spatially dynamic protein presentation that mimics the lateral diffusion of proteins in biological cell membranes. When evaluated in mouse models of melanoma, anisotropic aAPCs can reduce tumor growth and enhance survival. In addition, when PLGA aAPCs, that activate T cells, are combined with the checkpoint inhibitor anti-programmed death 1, that prevents downregulation of T cells, we find a synergistic effect and improved efficacy at treating melanoma. These biologically-inspired immunoengineered particles can engineer T cells in vivo and could be an enabling technology for cancer immunotherapy.

Quantitative, Molecular-Level Analysis of the Serum Antibody Repertoire Reveals Unanticipated Features of the Response to Influenza Vaccination

Conference: Translational Medicine and Bioengineering Conference

Year: 2016

Proceeding: Translational Medicine and Bioengineering Conference

Group: General Submissions

Session: Immunoengineering

Time: Saturday, November 12, 2016 - 2:15pm-2:30pm

Authors:

Lee, J., The University of Texas at Austin
Georgiou, G., The University of Texas at Austin
DeKosky, B. J., University of Kansas
Boutz, D. R., The University of Texas at Austin
Chromikova, V., Icahn School of Medicine at Mount Sinai
Joyce, M. G., National Institutes of Health
Vollmers, C., Stanford University
Leung, K., National Institutes of Health
Horton, A. P., The University of Texas at Austin
Lee, C. H., The University of Texas at Austin
Lavinder, J. J., The University of Texas at Austin
Murrin, E. M., University of Texas at Austin
Chrysostomou, C., The University of Texas at Austin
Hoi, K. H., The University of Texas at Austin
Park, D., University of Texas at Austin
Davis, M. M., Stanford University, Howard Hughes Medical Institute
Ellington, A. D., The University of Texas at Austin
Marcotte, E. M., The University of Texas at Austin
Ippolito, G. C., The University of Texas at Austin
Quake, S. R., Stanford University
Kwong, P. D., National Institutes of Health
Krammer, F., Icahn School of Medicine at Mount Sinai

Mascola, J. R., Vaccine Research Center, National Institutes of Health
Wilson, P. C., The University of Chicago
Ross, T. M., University of Georgia
Bloom, C. E., University of Georgia
Dekker, C. L., Stanford University
Popova, L. I., The University of Chicago
Kong, W. P., Vaccine Research Center, National Institutes of Health
Wang, L., Vaccine Research Center, National Institutes of Health
Zhang, Y., Vaccine Research Center, National Institutes of Health
Zhang, B., Vaccine Research Center, National Institutes of Health
Druz, A., Vaccine Research Center, National Institutes of Health
Thomas, P. V., Vaccine Research Center, National Institutes of Health
Tsybovsky, Y., Frederick National Laboratory for Cancer Research

Vaccination is the most effective means to protect populations against infectious diseases, conferring protection by eliciting a diverse repertoire of virus-specific antibodies that circulate in blood and afford immunity by binding and neutralizing viruses. For influenza, which is a respiratory virus that poses a constant threat of a pandemic, seasonal vaccination is the best prophylactics; however, it leaves a large fraction of the population vulnerable, having less than 60% efficacy in young adults and substantially lower for at-risk groups. To design better vaccination strategies aimed at increasing its effectiveness, a molecular-level understanding of serological immunity to influenza is needed, but it has been confounded by the complexity of the polyclonal antibody response. Here, we used novel high-resolution proteomics coupled with high-throughput single B-cell sequencing technologies to quantitatively identify the antibodies elicited by 2011-2012 seasonal flu vaccination in four healthy donors and subsequently determined the biochemical and functional properties of prevalent serum antibodies. Following vaccination, we determined that the boosting of the antibodies pre-existing prior to vaccination accounted for ~60% of the vaccine response, indicating the substantial degree of antibody response being biased by the antibodies that were already existing prior to vaccination. We also observed a surprisingly high fraction of antibodies binding to both of the two highly diverse virus strains (group 1 and group 2). Characterization of these cross-reactive antibodies revealed broad recognition of historical virus strains and led to the discovery of a new class of antibodies that protect mice against H1N1 and H3N2 virus lethal challenges. This class of antibodies recognized the same conserved region in the head domain of hemagglutinin that is exposed only in its monomeric form. This is the first-ever report providing an extensive molecular-level description of the identities, dynamics and functions of the monoclonal antibodies that comprise the antibody response to seasonal flu vaccination. Collectively, our data provide unprecedented insights on the serological responses to influenza vaccination that have direct implications for the engineering of immunogens for a future vaccine for influenza.

Synthetic Biology in Cancer Cellular Immunotherapy

Conference: Translational Medicine and Bioengineering Conference

Year: 2016

Proceeding: Translational Medicine and Bioengineering Conference

Group: General Submissions

Session: Immunoengineering

Time: Saturday, November 12, 2016 - 2:30pm-2:45pm

Authors:

Wong, W., Boston University

The transfer of tumor-targeting T cells to patients is a promising approach for cancer immunotherapy. In particular, chimeric antigen receptors (CARs) modified T cells had demonstrated phenomenal efficacy against lymphoid leukemia, with more than 90% complete remission being observed. Despite these encouraging results, many challenges remain to be addressed before T cell therapy can be widely adopted for other cancers. Limiting the toxicity without comprising their efficacy are some of the key hurdles that need to be overcome. Here I will describe our recent efforts in using synthetic biology to develop drug-controllable genetic circuits and signaling switches for temporally modulating T cell activation. Specifically, I will highlight a set of multi-state memory switches for regulating multiple CARs expression, which enable the tuning of CAR expression level and ligand specificity. In addition, I will also discuss our novel dual small-molecule-gated signaling switches for ON/OFF control (one drug for ON switch, one drug for OFF switch) of T cell activation originated from T cell receptors and CARs. These genetic tools will provide a flexible tool kit for managing or preventing the toxicity associated with T cell therapy.

Tumor Targeting and Immunotherapy Enabled By an Engineered Peptide-Fc Fusion

Conference: Translational Medicine and Bioengineering Conference

Year: 2016

Proceeding: Translational Medicine and Bioengineering Conference

Group: General Submissions

Session: Immunoengineering

Time: Saturday, November 12, 2016 - 1:00pm-1:30pm

Authors:

Cochran, J. R., Stanford University

In this talk, I will discuss the discovery and development of a novel therapeutic approach to targeted cancer treatment, spanningÂ a broad range of efforts in the areas of protein engineering, biochemical and biophysical analyses, and preclinical testing. This project has benefited significantly from multi-disciplinary expertise of scientists, engineers, and clinicians across several institutions, including Stanford, MIT, and Dana Farber; such highly collaborative science highlights a productive formula for addressing complex biomedical problems. In the spirit of the meeting theme of âtranslational medicine and bioengineeringâ, I will also discuss considerations for the successful development of protein biologics, including production yield, stability, therapeutic efficacy, toxicity, and immunogenicity.

Discovery and Engineering of Disulfide Bond Switches in Mammals

Conference: Translational Medicine and Bioengineering Conference

Year: 2016

Proceeding: Translational Medicine and Bioengineering Conference

Group: General Submissions

Session: Biomanufacturing

Time: Sunday, November 13, 2016 - 9:15am-9:45am

Authors:

Khosla, C., Stanford University

Due to the oxidizing nature of the extracellular environment, disulfide bonds found in secreted proteins were once believed to inert. The discovery of redox-sensitive disulfides that, once cleaved, could lead to changes in protein activity has challenged this notion. This lecture will summarize the fundamental biochemistry as well as the engineering opportunities associated with such functional disulfide bonds.

Next-Generation Sequencing and Omics Technologies for Biotherapeutic Manufacturing Development

Conference: Translational Medicine and Bioengineering Conference

Year: 2016

Proceeding: Translational Medicine and Bioengineering Conference

Group: General Submissions

Session: Biomanufacturing

Time: Sunday, November 13, 2016 - 10:00am-10:15am

Authors:

Groot, J., Biogen
Wright, C., Biogen
Davis, S., Biogen

Driven by a continuous innovation of experimental and computational methodologies, Next-Generation Sequencing (NGS) and other Omics technologies continue to advance at a rapid pace. These technologies have led to a revolution in biological research and medicine but have applications well beyond those areas. One field in which these technologies have started to make a positive impact is the development of cell lines and bioprocesses for biotherapeutic manufacturing.
At Biogen, we are using NGS to both characterize therapeutic production clones and evaluate manufacturing culture processes, many of which use Chines Hamster Ovary (CHO) cells. One application is an NGS screen of candidate production clones to ensure their biotherapeutic mRNA sequence is free of mutations. For this screen we developed a targeted mRNA sequencing assay that multiplexes samples of multiple clones onto a single sequencer lane. This mutation screen has become a standard part of our cell line development workflow to which it adds quality and rigor on decisions about candidate production clones. Another application uses targeted genome sequencing to characterize integration sites of the biotherapeutic transgene. This integration site assay allows for clone fingerprinting, provides information on genetic stability, and supports efforts for locus specific engineering of transgene integration into the CHO genome. In addition to clone characterization we use NGS (RNA-Seq) to obtain insights into the cell biology of bioprocessing. The genome for CHO is much less resolved than the genome for human or mouse and to set up a robust omics platform around CHO we evaluated

several draft CHO genome assemblies and annotations. In addition, we developed a standardized computational pipeline to process CHO RNA-Seq data. This enabled comparing gene expression profiles across our bioprocess which led to the identification of changes related to particular growth stages and transgene expression.

In conclusion we successfully developed and implemented a toolbox of NGS technologies that we continue to refine and expand. These NGS tools provide improvements to our cell line development workflows and generate new insights into our CHO production processes

Recent Developments in Downstream Bioprocessing: Multimodal Chromatography, Affinity Precipitation and Integrated Bioprocessing

Conference: Translational Medicine and Bioengineering Conference

Year: 2016

Proceeding: Translational Medicine and Bioengineering Conference

Group: General Submissions

Session: Biomanufacturing

Time: Sunday, November 13, 2016 - 10:45am-11:00am

Authors:

Cramer, S., RPI

This presentation will focus on recent developments in our lab in three areas of downstream bioprocessing; multimodal chromatography, affinity precipitation and integrated bioprocessing. Fundamental studies into the nature of selectivity in multimodal chromatography will first be presented. This work will include chromatographic studies with various protein and multimodal ligand libraries and novel predictive approaches including course grained protein binding maps and quantitative structure activity relationship (QSAR) models using new classes protein molecular descriptors. Work will then be presented on affinity precipitation using smart biopolymers for the simultaneous recovery and purification of biological products. Finally, some results are presented on the development of integrated biomanufacturing systems using both affinity peptide capture as well as an optimized series of multimodal chromatographic operations.

Stem Cell-Based Product Biomanufacturing: From Cultivation Modalities to Population Based-Models

Conference: Translational Medicine and Bioengineering Conference

Year: 2016

Proceeding: Translational Medicine and Bioengineering Conference

Group: General Submissions

Session: Biomanufacturing

Time: Sunday, November 13, 2016 - 10:15am-10:30am

Authors:

Tzanakakis, E. S., Tufts University

The engineering of bioprocesses for stem-cell based therapy products can benefit from the implementation of the quality by design paradigm in the manufacturing of biopharmaceutics. This approach requires better understanding of process parameters and product characteristics. The list of involved tasks is extensive but this presentation will focus on three core aspects. First, different cultivation modes are available for stem cell expansion and differentiation. Stirred suspension bioreactors, which are commonly used for high-density culture of vaccine and biosimilar producer cells, allow the propagation of stem cells as aggregates, on microcarriers or after encapsulation affording process design flexibility. Selection of the culture modality is largely dictated by attributes of the final cell type(s) including sensitivity to shear, and dependence of commitment on concentration gradients and paracrine signals. Second, cell growth in these systems is intimately influenced by the formulation of media and surfaces shaping the culture environment. After years of intense research efforts, the development is coming to fruition of chemically defined, xeno-free media and substrates conforming to current good manufacturing practices (cGMP) for pluripotent cell cultivation and directed differentiation along specific trajectories. The increasing reliance on small molecules to replace growth factors and cytokines has the benefits of reduced cost and increased medium stability. Third, stem cell bioprocess engineering is expected to gain significantly from mathematical models capturing the heterogeneity of stem/progenitor cell ensembles. Population-based models with multiple process variables as inputs quantified via process analytical technology can be used to map and control process (e.g. differentiation strength) or product attributes (cell phenotype or function) within a given design space. These models can be combined with next-generation sequencing and proteomic data to gain insight into ways by which gene/protein networks interact with the process environment and guide optimization efforts. Despite the significant strides noted in the areas discussed here, integration of approaches from a broad range of disciplines will be essential for the realization of robust, economical processes for stem cell-based product manufacturing.

Sugar and Spice: Systems Biotechnology Approaches to Interpret and Improve Mammalian Cell Factories

Conference: Translational Medicine and Bioengineering Conference

Year: 2016

Proceeding: Translational Medicine and Bioengineering Conference

Group: General Submissions

Session: Biomanufacturing

Time: Sunday, November 13, 2016 - 10:30am-10:45am

Authors:

Betenbaugh, M. J., Johns Hopkins University

Recombinant therapeutics are often complex proteins that require multiple post-translational modifications including glycosylation and protein assembly among other changes. Their biological function and in vivo circulatory activity can depend intimately on these processing events, often occurring within the secretory compartments of the producer cells. Most importantly, a number of factors can impact glycosylation in Chinese hamster ovary and other cell lines. Therefore, the capacity to both elucidate and manipulate these modifications will be useful going forward. In this study, engineering and analytic methods are being applied to both determine the glycosylation structures and if needed change these glycoforms. One approach being used to is to alter the cell lines in order to alter the glycosylation capabilities of CHO cells. Another alternative is to alter the media such that the glycosylation patterns change. It is equally important to be able to analyze the effect of these engineering modifications and thus analytical tools must be implemented to evaluate the glycan patterns on proteins emerging from these hosts. Finally, it is also important to understand the impact of changes to the cell processing capabilities and mathematical modeling tools can be helpful both for understanding how these processing events occur and controlling them for a desired outcome. Such a comprehensive systems approach will enhance our ability to control mammalian glycosylation and other cellular events and in turn lead to improved qualities for target biopharmaceuticals in the future.

Development of a CHO-Based Cell-Free Biomanufacturing Platform for Synthesis of Active Human Monoclonal Antibodies

Conference: Translational Medicine and Bioengineering Conference

Year: 2016

Proceeding: Translational Medicine and Bioengineering Conference

Group: General Submissions

Session: Biomanufacturing

Time: Sunday, November 13, 2016 - 9:45am-10:00am

Authors:

Martin, R. W., Northwestern University
Majewska, N.
Chen, C.
Jewett, M. C., Northwestern University
Roy, V., MedImmune Inc.

Chinese Hamster Ovary (CHO) cells are routinely optimized to stably express monoclonal antibodies (mAbs) at high titers. However, CHO cell technology can be unwieldy for rapid, high-throughput production and screening of candidate therapeutic proteins. Here, we have developed a cell-free protein biomanufacturing platform utilizing a commercially available CHO extract for the cell-free synthesis of mAbs. Our specific aims were to: (1) optimize cell-free reaction parameters to maximize protein yields, (2) screen additives to enhance disulfide bond formation for intact mAb production, and (3) establish the extract as a predictable tool for ranking yields of candidate antibodies. Optimization of the reaction conditions increased protein yields 8-fold to 700 mg/L using green fluorescent protein (GFP) as a reporter. Additives such as oxidized glutathione established an oxidizing cell-free environment for disulfide bond formation, and chaperones with disulfide isomerase activity further increased intact mAb yields 2-fold. Lastly, light and heavy chain expression was optimized to minimize aggregation products. Using our optimized platform, we demonstrate for the first time the cell-free synthesis of active, intact mAb using CHO extract. Additionally, the cell-free system demonstrated the capacity to rank known high and low producing mAb sequences in the same order as seen from CHO cell-based platforms. Unlike stable or transient transfection-based screening, which requires 8-14 days for setup and execution, results using our CHO-based CFPS system are attained within 2 days.

Further development of the customizable cell-free platform would provide a tool for rapid, high-throughput ranking of mAb producers and has implications for the synthesis of difficult-to-express proteins.

De Novo Prediction of Human Reprogramming Using Nuclear Imaging Data

Conference: Translational Medicine and Bioengineering Conference

Year: 2016

Proceeding: Translational Medicine and Bioengineering Conference

Group: General Submissions

Session: Stem Cells/ Regenerative Medicine

Time: Sunday, November 13, 2016 - 4:00pm-4:15pm

Authors:

Saha, K., Whitehead Institute for Biomedical Research
Harkness, T., University of Wisconsin-Madison
Carlson-Stevermer, J., University of Wisconsin-Madison
Prestil, R., University of Wisconsin-Madison
Seymour, S., University of Wisconsin-Madison
Abdeen, A., University of Illinois at Urbana-Champaign

Molecular mechanisms of reprogramming terminally differentiated cells to a pluripotent state still are poorly understood. As a result, standard reprogramming techniques are still noisy and inefficient. To address these shortcomings, we have developed a novel platform that allows for the dynamic tracking of subpopulations in a longitudinal manner across long periods of time. This platform is a simple approach that combines live-cell microscopy with surface-modified multiwell plates that separates thousands of cell populations. With this we are able to both watch and physically constrain cells into discrete islands during reprogramming. By watching subpopulations in real time we are able to distinguish intermediate states that either contribute or are detrimental to complete reprogramming. Furthermore by controlling the island geometry at the microscale, we are able to selectively activate mechanotransduction pathways (e.g., YAP/Taz) to promote transition through the endothelial-mesenchymal transition during the reprogramming process. Direct manipulation of the YAP/Taz pathway via lentiviral overexpression also affected the reprogramming process. Overall, this simple platform allows us to expand our understanding of intermediate cell states and increasing the overall efficiency of the reprogramming process.

Defined, Xeno-Free Platforms for Manufacturing Cardiovascular Cell Types from Human Pluripotent Stem Cells

Conference: Translational Medicine and Bioengineering Conference

Year: 2016

Proceeding: Translational Medicine and Bioengineering Conference

Group: General Submissions

Session: Stem Cells/ Regenerative Medicine

Time: Sunday, November 13, 2016 - 3:00pm-3:30pm

Authors:

Palecek, S., University of Wisconsin

Over the past decade tremendous progress has been made in developing processes to direct the differentiation of human pluripotent stem cells to various specialized cell types for a variety of applications including development and disease modeling, drug screening and evaluation, and regenerative therapeutics. Typically these approaches involve construction of biomimetic microenvironments to guide stem cell differentiation in a developmentally-relevant manner. However, these approaches are often costly, inefficient, difficult to scale, and not compatible with cGMP manufacturing. The next generation of differentiation platforms requires simple, robust, inexpensive, and xeno-free characteristics. In this presentation I will discuss evolution of differentiation platforms for cardiovascular. We have identified canonical Wnt signaling as a key regulator of hPSC differentiation to cardiomyocyte and designed a protocol that produces high purity cardiomyocytes in a defined, xeno-free, growth factor-free system via appropriate temporal presentation of small molecule modulators of Wnt signaling. Furthermore, we have determined that a different canonical Wnt signaling profiles can direct hPSCs to vascular endothelial progenitors and epicardial cells in a defined, growth factor-free differentiation process. These epicardial cells can give rise to cardiac fibroblasts and smooth muscle cells upon induction of an epithelial-to-mesenchymal transition. Thus, by stage-specific modulation of canonical Wnt signaling and other developmental pathways we can generate a diverse array hPSC-derived cardiac cells in defined, xeno-free conditions for in vitro studies, drug screening and toxicology analyses, and development of regenerative therapies.

Distinct and Shared Determinants Impacting Cardiomyocyte Contractility in Multi-Lineage Competent Ethnically Diverse iPSCs

Conference: Translational Medicine and Bioengineering Conference

Year: 2016

Proceeding: Translational Medicine and Bioengineering Conference

Group: General Submissions

Session: Stem Cells/ Regenerative Medicine

Time: Sunday, November 13, 2016 - 3:45pm-4:00pm

Authors:

Tomov, M., SUNY Polytechnic Institute
Paluh, J. L., SUNY-CNSE
Olmsted, Z. T., SUNY-CNSE
Otu, H.
Buck, M., SUNY Buffalo
Dogan, H.
Cibelli, J.

The promise of stem cell based therapeutic intervention in heart disease holds hope to improve long-term therapeutic outcomes, but is not yet realized. Strategies with bone marrow derived stem cells have been prevalent in clinical trials however other stem cell resources may offer advantages. The ability to use reprogrammed somatic cells for patient-optimized therapies, either from direct reprogramming to cardiac lineage or reprogramming first to pluripotent stem cells, requires expanding our understanding of how the epigenetic landscape affects downstream differentiation into cardiomyocytes or other specialized cardiac cells. Population diversity, including ethnicity should be included as resources become available. We derived multiple replicate iPSC lines from apparently healthy African American, Hispanic-Latino and Asian self-designated ethnically diverse (ED) origins and verified normal karyotype, teratoma formation, pluripotency biomarkers, and tri-lineage in vitro commitment. In multi-lineage derivation of cells of therapeutic interest we observed differences between replicate lines in efficiency to generate cardiomyocytes although similar and robust differentiation to multiple neural, pancreatic, and smooth muscle cell types. By bioinformatics of RNA-Seq and ChIP-seq pluripotency data sets for two replicate Asian and Hispanic-Latino ED-iPSC lines our analysis identified shared and distinct genes and contributing pathways in the replicate ED-iPSC lines. We further compared our analysis to larger genome wide association studies on ethnic populations to enhance our ability to understand how reprogramming to iPSC impacts genes and pathways contributing to cardiomyocyte contractility potential. In our analysis of cardiomyocytes we utilized custom lithography templated microarrays for quantitative assessment of multiple phenotypes. Smaller, yet data-rich analyses of ethnically diverse cell sets such as the data we generated on cardiomyocytes from ethnically diverse individuals, the first of its kind, are an important step and complement larger GWAS studies. In addition, in vitro models more readily provide genetic replicates versus the more complex phenotyping of large patient populations. The current study and data represent a new resource for continued investigation by the stem cell biomedical community and may find additional usefulness in bioengineered platforms for drug testing and metabolomics using ethnically diverse lines.

Engineered Single Human iPSC-Cardiomyocytes to Assay Contractile Defects Induced By Drugs and Disease States

Conference: Translational Medicine and Bioengineering Conference

Year: 2016

Proceeding: Translational Medicine and Bioengineering Conference

Group: General Submissions

Session: Stem Cells/ Regenerative Medicine

Time: Sunday, November 13, 2016 - 3:30pm-3:45pm

Authors:

Ribeiro, A. J. S., Stanford University
Ang, Y. S., University of California San Francisco
Mandegar, M.
Conklin, B., Gladstone Institute of Cardiovascular Disease
Srivastava, D., Gladstone Institute of Cardiovascular Disease
Pruitt, B. L., Stanford University

Human induced pluripotent stem cells (iPSCs) can be differentiated into cardiomyocytes (hiPSC-CMs) and have the potential to model cardiac function. We developed, studied and validated a platform where hiPSC-CMs are engineered to present a more mature structural organization and are also functionally assayed. We specifically analyzed different parameters related to the contractile function of these engineered hiPSC-CMs and tested their ability to model the effects of drugs and disease states in cardiac function.

Introduction: The contractility of cardiomyocytes enables the beating of the heart. Sarcomeres are the contractile units of cardiomyocytes. In mature cardiomyocytes sarcomeres are distributed in series along myofibrils that are well aligned in the intracellular space. Within each sarcomere, the contractile interactions of myosins with actin filaments induce sarcomere shortening and cell beating. However, hiPSC-CMs lack the distribution and organization of sarcomeres and myofibrils that characterize mature cardiomyocytes. We engineered single hiPSC-CMs on polyacrylamide hydrogels with physiological stiffness to induce a mature cell shape and more mature organization of sarcomeres and myofibrils. We tested how the contractility of engineered hiPSC-CMs varied under disease conditions and due to drugs known to affect heart contractile activity. We also investigated the role of extracellular physical cues in driving the structural maturity in engineered hiPSC-CMs via a tension-driven mechanism.

Materials and Methods:To engineer single hiPSC-CMs, we seeded cells singularized from beating monolayers on polyacrylamide substrates of physiological stiffnesses with micropatterned rectangles composed of Matrigel. Rectangular micropatterns in association with a substrate stiffness of 10 kPa drove hiPSC-CMs to have a more mature shape and an aligned organization of sarcomeres and myofibrils. We dispersed fluorescent microbeads in the polyacrylamide substrate to calculate contractile and kinetic parameters of cell beating from traction force microscopy measurements. We also labeled actin in sarcomeres with LifeAct in live cells to measure the dynamics of sarcomere shortening and myofibril organization. We tested the role of tension in cell function by varying substrate stiffness and the length/width ratio of micropatterns. We analyzed variations in contractility and sarcomere activity induced by mutations that model heart disease, such as MYBPC3 gene knockout and mutations in GATA4 and MYH7 genes. Contractile variations induced by incubating cells in caffeine, isoproterenol and omecamtiv mercanbil were also studied.

Results and Discussion: We observed that cell shape and the stiffness of the substrate induced structural maturation of sarcomeres and myofibrils in engineered hiPSC-CMs via a tension mechanism. We detected different variations in the contractility of cells due to disease states and induced by isoproterenol and omecamtiv mercanbil.

Conclusions and Future Work: We validated a platform for systematic measuring the contractility of hiPSC-CMs and modeling cardiac activity in a dish to study the effects of heart disease and drugs in a dish. We also identified intracellular tension as a key cue in the structural maturation of these engineered cells. Altering intracellular and extracellular biological properties may further control tension in structurally maturing hiPSC-CMs.

Migration of Human Retinal Progenitor Cells in 3D Collagen Gels

Conference: Translational Medicine and Bioengineering Conference

Year: 2016

Proceeding: Translational Medicine and Bioengineering Conference

Group: General Submissions

Session: Stem Cells/ Regenerative Medicine

Time: Sunday, November 13, 2016 - 4:45pm-5:00pm

Authors:

Carrier, R. L., Northeastern University
Kundu, J.
Blanton, D.
Baranov, P.
Oudin, M.
Young, M.
Miller, M.
Lauffenburger, D., MIT

Migration of Human Retinal Progenitor Cells in 3D Collagen Gels

Joydip Kundu[1,2], Douglas Blanton[1], Petr Baranov[2], Madeleine J Oudin[3], Michael J. Young[2],

Miles A. Miller[4], Douglas A Lauffenburger[4], Rebecca L. Carrier[1]

[1] Chemical Engineering, Northeastern University, Boston, MA 02115

[2] Schepens Eye Research Institute, an affiliate of Harvard Medical School, Boston, MA 02114

[3] David Koch Institute for Integrative Cancer Research, Massachusetts Institute of Technology, Cambridge, MA 02115

[4] Bioengineering Department, Massachusetts Institute of Technology, Cambridge, MA 02115

Abstract

Introduction: Retinal degeneration is the leading cause of blindness worldwide. It is currently estimated that 25 to 30 million people worldwide suffer from age-related macular degeneration, with no known cure available to date. Subretinal implantation of retinal progenitor cells has shown great promise in models of retinal degeneration for restoration of vision, but is limited by extremely low integration (<2%) into the retina. Understanding the cues required to promote cell behaviors essential to successful integration (e.g., migration) will enable rational design of cell delivery vehicles to enhance integration within host retina following transplantation. Objective: Investigate the impact of matrix stiffness and soluble cues (varied growth factor concentrations) on human retinal progenitor cells (hRPCs) migration. Methodology: In this study, the migration of hRPCs within thermo-responsive collagen gels of varying concentrations (3mg/ml, 2mg/ml and 1mg/ml) was analyzed using a high-throughput 3D migration assay. The storage moduli (stiffness) of the collagen gels (3mg/ml, 2mg/ml and 1mg/ml) were determined using a rheometer. The impact of addition of growth factors known to be significant in retinal development, retinal regeneration in non-mammals, and neuronal cell migration: epidermal growth factor (EGF) and fibroblastic growth factor (FGF), was studied using hRPCs in 2mg/ml collagen gels. Results: The mean migration distances of the encapsulated hRPCs within gels after 48 hours were 39±3µm (3mg/ml collagen), 46±3µm (2mg/ml collagen) and 45±5µm (1mg/ml collagen). The storage modulus (measured at 10% strain) of 3mg/ml collagen gels (1.29 Pa) was 2-fold higher than that of 2mg/ml (0.64 Pa) and 1mg/ml (0.50 Pa) collagen gels. The gel stiffness did not have a significant impact on the migration of the retinal progenitors. The addition of EGF and FGF (at concentrations 10ng/ml and 100ng/ml) in 2mg/ml collagen gel increased hRPC migration by 2.2-2.3 fold compared to media alone. Cells exposed to 100ng/ml of EGF and FGF showed 1.5-1.6 fold increased migration compared to those treated with 10ng/ml of these cytokines. The impact of cytokines on cell signaling events in hRPCs is being explored using multiplex cytokine analysis to elucidate the signaling pathways critical for hRPC migration. Conclusion: These results motivate further exploration of the impact of soluble cues in the context of different hydrogels, representative of either the retinal microenvironment or potential cell delivery vehicles, on hRPC migration to inform rational design of delivery vehicles enabling successful cell transplantation to treat retinal degeneration.

Acknowledgements: This study was supported by Northeastern University (Tier 1 Provost Grant) and NIH Grant 1R21EY021312 (NEI).

Utilizing Developmental Cues to 'Engineer' Native Tissues from Human Induced Pluripotent Stem Cells (iPSC)

Conference: Translational Medicine and Bioengineering Conference

Year: 2016

Proceeding: Translational Medicine and Bioengineering Conference

Group: General Submissions

Session: Stem Cells/ Regenerative Medicine

Time: Sunday, November 13, 2016 - 4:30pm-4:45pm

Authors:

Bhutani, N., Stanford University
Lee, J., Stanford University

The remarkable discovery of induced pluripotent stem cells (iPSC), generated by âreprogramminga of somatic cells by introducing a few transcription factors, has provided an unprecedented source of patient-specific cells for regenerative medicine. Besides pluripotency, somatic cells have also been demonstrated to âreprogramâ or âtransdifferentiateâ directly into another somatic cell type. A repertoire of these reprogramming tools can therefore be utilized to âdesignâ and âengineerâ autologous, native tissues for defects and injuries. A challenge with iPSC-derived cells has been the variable developmental maturity of the cells, for example for muscle and heart tissues wherein the iPSC-derived cells are immature and not fully functional as in the native tissues. Developmental immaturity or pliability can however be advantageous for other tissues like cartilage that do not regenerate efficiently in adulthood thereby frequently leading to degenerative diseases like Osteoarthritis (OA). Cell-based therapeutic approaches for repairing focal cartilage defects have utilized autologous adult cartilage cells i.e. chondrocytes or adult mesenchymal stem cells (MSC) but with limited success due to generation of inferior fibrocartilage and paucity of cells. Indeed, recent reports have provided compelling evidence that juvenile chondrocytes (from donors below 13 years of age) are more efficient at generating cartilage tissue as compared to adult chondrocytes. Our goal has been to identify the molecular basis for such a superior regenerative capability and to test whether iPSC-derived chondrocytes mimic the juvenile chondrocytes in a molecular and functional manner. Towards this goal, we firstly developed a defined growth factor based protocol for inducing chondrogenesis in human iPSC with high efficiency (Lee et al, FASEB J, 2015). A major advantage of this method is that the human iPSC are efficiently differentiated into chondrocytes with an articular rather than fibrocartilage-like phenotype. Molecular characterization of the differentiating human iPSC revealed that differentiation proceeds through an intermediate $Sox9^{low}$ $CD44^{low}$ mesodermal population leading to a fairly homogenous $Sox9^{high}$ $CD140^{high}$ $CD44^{high}$ hiChondrocyte population expressing high level of the articular genes, Wnt9a and SOSTDC1, chondrogenic transcription factors (Sox 5/6/9) and Col2a1, but not fibrocartilage marker Col1a1 or the hypertrophic marker Col10. The iPSC-derived chondrocytes (hichondrocytes) are stable and comparable to adult chondrocytes in their ability to generate cartilage tissue. In addition, hichondrocytes mimic juvenile chondrocytes in faster cell proliferation than adult chondrocytes, providing an advantage in terms of cell expansion and tissue generation. Additionally, in vivocartilage generation in mice by the hichondrocytes generated strictly cartilage and not teratomas demonstrating the efficacy and safety of the hiChondrocytes. Whole genome transcriptome analyses revealed a subset of factors conserved between juvenile and iPSC-derived chondrocytes that are likely responsible for their superior regenerative potential. These studies have identified novel factors that enhance stem cell-mediated cartilage regeneration and highlight the importance of studying developmental processes to understand and identify the ideal cells to engineer to realize the generation of a functional native tissue.

Vascular Differentiation and Engineering from Human Pluripotent Stem Cells

Conference: Translational Medicine and Bioengineering Conference

Year: 2016

Proceeding: Translational Medicine and Bioengineering Conference

Group: General Submissions

Session: Stem Cells/ Regenerative Medicine

Time: Sunday, November 13, 2016 - 4:15pm-4:30pm

Authors:

Gerecht, S., Johns Hopkins University

Proper blood vessel networks are necessary for constructing and re-constructing tissues, promoting wound healing, and delivering metabolic necessities throughout the body. Since their advent, human induced pluripotent stem cells have played a growing role in tissue engineering for building or reconstructing functional human tissue that is also patient specific. In this talk, Iâll present our efforts to derive vascular cells including endothelial, pericytes and smooth muscle cells from pluripotent stem cells, towards the formation of functional networks. Focus will be given on the effect of physicochemical cues on cellular fate, maturation and tissue assembly.

Analyzing and Predicting Antibody Responses

Conference: Translational Medicine and Bioengineering Conference

Year: 2016

Proceeding: Translational Medicine and Bioengineering Conference

Group: Poster Submissions

Session: Poster Session

Time: Saturday, November 12, 2016 - 5:00pm-7:00pm

Authors:

Reddy, S. T., University of Texas Austin

The ability to predict and correspondingly manipulate an adaptive immune response would be highly valuable for biotechnology and medicine. To achieve this requires a greater molecular understanding of adaptive immunity. Recent advances in high-throughput immunoglobulin (antibody) repertoire sequencing (Ig-seq) are enabling highly quantitative analysis of adaptive immune responses. This increased immunological insight has been applied to fields as varied as lymphocyte development and differentiation, immunodiagnostics discovery, vaccine development, cancer immunotherapy, and monoclonal antibody discovery. Here I will present, the computational and experimental methods my group has developed in Ig-seq and their associated applications. For, example we have recently developed a comprehensive error and bias correction method that enables highly accurate Ig-seq. We are also applying machine-learning approaches that enable prediction of the immune status of individual antibody clones. Finally, in the context of adaptive immunity, we addressed the age-old question of the balance between nature, nurture and noise. It has remained unclear to what extent nature (genetic background) and nurture (antigen challenge) predetermine an antibody response and what fraction is private to an individual (noise). By employing ultra-deep Ig-seq (400 million reads) in mice, we discovered that that nature and nurture dictate up to 99% of the development of immunoglobulin repertoires. This is in contrast to current immunological dogma, which suggests that an antibody response is overwhelmingly unpredictable. The resolution of this question resolves a decisive blank area in the understanding and prediction of adaptive immunity.

Aptamer Micelles Targeting a Novel Chemokine for Cancer Therapeutics

Conference: Translational Medicine and Bioengineering Conference

Year: 2016

Proceeding: Translational Medicine and Bioengineering Conference

Group: Poster Submissions

Session: Poster Session

Time: Saturday, November 12, 2016 - 5:00pm-7:00pm

Authors:

Kokkoli, E., University of Minnesota
Harris, M. A., University of Minnesota
Pearce, T. R., University of Minnesota
Pengo, T., University of Minnesota
Kuang, H., University of Minnesota
Forster, C. L., University of Minnesota

There is a growing need for cancer therapeutics to be delivered without off-target side effects; in order to do this the cancer cells must express a unique target. We have previously used systematic evolution of ligands by exponential enrichment (SELEX) to develop a single stranded DNA aptamer, FKN-S2, which binds to a novel extracellular chemokine, fractalkine. Unlike other chemokines, fractalkine has been shown to act as an adhesion molecule and is expressed on inflamed and cancerous cells thus making it a promising target for cancer therapeutics. We have synthesized FKN-S2 as an amphiphile by conjugating it to a C16 dialkyl tail and showed that the FKN-S2 aptamer-amphiphiles self-assemble into micelles. This work investigates for the first time the extent and efficacy of using the FKN-S2 aptamer micelles as a targeted delivery vehicle for in vitro and in vivo delivery to fractalkine-expressing mouse colon adenocarcinoma cells. Our in vitro cell binding and trafficking results show that FKN-S2 micelles (functionalized with polyethylene glycol (PEG) or non-PEGylated) can bind and internalize to fractalkine-expressing cancer cells with minimal surface delivery to healthy cells. In addition, our in vivo microPET/CT imaging and biodistribution studies and ex-vivo biodistribution and confocal imaging studies demonstrate that even though both PEGylated FKN-S2 and control micelles can accumulate to the tumor site only the PEGylated FKN-S2 micelles internalize to fractalkine-expressing tumor cancer cells, thus demonstrating the potential of our approach and showing that fractalkine may serve as a specific target for nanoparticle delivery to cancer cells.

Biophysical Cues Modulate the Anti-Fibrotic Response of Rabbit Corneal Stromal Cells

Conference: Translational Medicine and Bioengineering Conference

Year: 2016

Proceeding: Translational Medicine and Bioengineering Conference

Group: Poster Submissions

Session: Poster Session

Time: Saturday, November 12, 2016 - 5:00pm-7:00pm

Authors:

Song, Y.
Rewinski, E.
Murphy, C., School of Medicine, University of California Davis
Raghunathan, V.
Thomasy, S.

During corneal wound healing, changes to the cellular microenvironment involve alterations in the expression of soluble signaling molecules [e.g. transforming growth factor (TGFÎ²)] as well as modulation of the biophysical attributes of the corneal stroma. These stimuli mediate the transformation of quiescent Keratocytes to activated Fibroblasts and further to contractile Myofibroblasts (KFM transition). Although, myofibroblast cells are critical for effective wound repair, excessive numbers and/or prolonged residence of these cells in the cornea can lead to stromal haze and fibrosis. Heat Shock Protein 90 (HSP90), a chaperone protein, stabilizes the TGFß1 pathway that is critical in KFM transformation. We hypothesized that inhibiting HSP90 would prevent the fibrotic phenotype of rabbit corneal stromal cells in vitro. This hypothesis further implies that HSP90 inhibition may minimize stromal haze in vivo. That biophysical stimuli such as topography and stiffness can mediate phenotypic alterations of corneal stromal cells is well established. However, the effect of biophysical parameters on governing the effect of drugs is lesser known. Here, primary rabbit corneal stromal fibroblasts (RCF) were cultured on both topographically patterned substrates [planar control and alternating ridges and grooves of 1400 nm pitch stamped into polyurethane (NOA81); pitch = groove + ridge width] as well as substrates having different stiffness [Tissue Culture Plastic (TCP; ~1GPa), as well polyacrylamide gels of 5kPa and 25 kPa moduli]. These cells were treated in the presence and/or absence of 10 ng/mL TGFß1, and 500 nM 17-(Allylamino)-17-demethoxygeldanamycin

(17-AAG; an Hsp90 inhibitor) for 72 hours. Protein and gene expression were quantified by Western blotting and qPCR, respectively. Immunocytochemistry was also used to visualize a myofibroblast marker, alpha smooth muscle actin (Î±SMA), of treated RCF cells. On planar NOA81 surfaces and on stiff TCP surfaces, TGFÃ1 treatment promoted the myofibroblast phenotype while 17-AAG promoted conversion of activated fibroblasts to the keratocyte phenotype, in the presence or absence of TGFÃ1. Interestingly, 25 kPa and 1400 nm pitch patterned substrates inhibited development of the myofibroblast phenotype in the presence or absence of TGFÃ1. However, for 17-AAG-treated cells, both 5 kPa and 1400 nm pitch substrates promoted development of the myofibroblast phenotype. Our data suggest that HSP90 inhibition may improve corneal wound healing outcomes but could have adverse consequences in normal unwounded corneas.

Characterization of an Electrostatic-Based Air Sampler for the Capture of Airborne Norovirus

Conference: Translational Medicine and Bioengineering Conference

Year: 2016

Proceeding: Translational Medicine and Bioengineering Conference

Group: Poster Submissions

Session: Poster Session

Time: Saturday, November 12, 2016 - 5:00pm-7:00pm

Authors:

Vega, D., Louis Stokes Alliance for Minority Participation

The need for an air sampler has been in demand, as airborne transmitted viruses like Norovirus, also known as the winter-vomiting disease, produces 685 million cases a year around the world1. The principal of electrostatic precipitation (ESP), although well-known for air filtration, has been recently used as a novel method for sampling for airborne viruses. The ESSA (ElectroStatic Sampler for Air) is a novel point-of-care device developed in-house for the capture of airborne virus. Two designs were created for the ESSA, that vary the needle organization; the cross design and the circular design. This project focuses on evaluating the performance of these two ESSA design for capturing airborne viruses. Based on the characterization results, the circular ESSA has the highest particle count efficiency while the cross ESSA has the best current versus voltage relationship.

Collagen Turnover-Stimulated Gene Delivery to Enhance Chronic Wound Repair

Conference: Translational Medicine and Bioengineering Conference

Year: 2016

Proceeding: Translational Medicine and Bioengineering Conference

Group: Poster Submissions

Session: Poster Session

Time: Saturday, November 12, 2016 - 5:00pm-7:00pm

Authors:

Urello, M. A., University of Delaware
Kiick, K. L., University of Delaware
Sullivan, M., University of Delaware

Biomaterial scaffolds are commonly employed to guide tissue regeneration. Ideal scaffolds act as analogs of the extracellular matrix, providing a cell-responsive structural support for cellular attachment and phenotypic commitment. For this purpose, the stable incorporation and controllable delivery of bioactive cues is required. In particular, the use of gene-containing biomaterials has been shown to address key obstacles in repair by enabling in situproduction of growth factors and other therapeutic molecules to modulate cell behavior. However, significant concerns with off-target delivery/safety and limited gene delivery efficacy have restricted clinical success. To address these issues, we have developed an innovative strategy for improving spatiotemporal control over gene expression by developing collagen-mimetic peptides (CMPs) to engineer DNA polyplex-modified collagens with tailorable release profiles and improved gene transfer. The unique collagen-like properties of CMPs permits their use as both adjustable tethers, via CMP-collagen hybridization, and as endocytic ligands.
For instance, variation of CMP-display on polyplex permitted tailoring of polyplex retention vs. release over a month, while non-modified polyplexes were retained for under 20 days. In cell culture studies, bound CMP-modified polyplexes also exhibited improved stability in the presence of serum-supplemented media over a 14-day period and different intracellular trafficking resulting in improved gene delivery efficiency. MMP-stimulated cells exhibited substantially higher levels of expression suggesting collagen remodeling mediates transfection. Additional experiments in a murine model demonstrated CMP-display enabled up to a 10-fold increase in transgene expression duration with average expression levels up to 2-orders of

magnitude greater than when non-modified polyplex were delivered. Furthermore, DNA polyplex-modified collagens encoding for platelet-derived growth factor -BB (PDGF-BB) exhibited the capacity to express enhanced levels of PDGF-BB and to subsequently encourage fibroblast proliferation/chemotaxis and wound closure when applied to an in vitro wound model. This work is the first to investigate the potential of exploiting the natural process of collagen remodeling to achieve enhanced gene delivery efficacy utilizing CMPs. These results suggest that this versatile approach has the potential to better integrate gene delivery with natural tissue repair while also improving polyplex safety and activity.

Combination Chemo- and Immunotherapy Against Latent Pulmonary Tuberculosis

Conference: Translational Medicine and Bioengineering Conference

Year: 2016

Proceeding: Translational Medicine and Bioengineering Conference

Group: Poster Submissions

Session: Poster Session

Time: Saturday, November 12, 2016 - 5:00pm-7:00pm

Authors:

Chow, G. I., University of Nevada, Reno
Kunda, N. K., University of New Mexico
Price, D. N., University of New Mexico
Stephens, L.
Muttil, P.

Annually, two million people die from active tuberculosis (TB). A significant fraction of the deaths is from the reactivation of a dormant form of tuberculosis called latent tuberculosis (LTB). LTB is more difficult to treat because LTB patients are asymptomatic and thus challenging to diagnose. Within the lungs of LTB patients, Mycobacterium tuberculosis (Mtb) is encapsulated by structures called granulomas which are impenetrable to anti-TB drugs when delivered by the conventional oral or injectable route. The goal of this study is to formulate a spray dried powder (SDP) that incorporates a live bacterial immunotherapeutic agent, bacillus CalmetteâGuÃ©rin (BCG), with an anti-TB drug, isoniazid, loaded in nanoparticles (INH NP). We hypothesize that when the SDP is delivered by the pulmonary route it will disrupt the lung granulomas leading to the elimination of Mtb from the LTB patient.
The INH containing poly (lactic-co-glycolic acid) (PLGA) NPs were synthesized by an emulsification/solvent evaporation method. Afterwards, the INH NPs were characterized for size, zeta-potential (Malvern Zetasizer), and INH drug loading in the NPs. INH loading in the NPs was assessed by UV spectrometry at an absorbance of 262 nm, measuring INH presence in the supernatant to indirectly calculate drug loading. The INH NPs were resuspended with BCG in a L-leucine solution at 1:1.5/ NPs: leucine ratio. Leucine was chosen as an excipient for its excellent flow properties. The suspension was spray-dried using a Büchi mini spray dryer at a flow rate of 3 mL/min, inlet temperature of 130±2 °C, and an outlet temperature of 48±2 °C.

The final mass of the SDP was used to calculate the final product yield. The particle size of the SDP was obtained using a Malvern Aerosizer.

The size of the INH NPs were 282.11±3.97 nm with a zeta-potential of -5.02±0.13 mV, making them ideal for inducing an immune response and maximizing the uptake by macrophages in the lungs. In addition, INH NPs were optimally loaded with 38.6±2.7 mg INH per 100 mg NPs, resulting in a SDP loaded with adequate INH to potentially eliminate the LTB in the lungs. The BCG loading in the SDP was 3.3 x 10^7 CFU/mg. We predict this BCG loading to be sufficient to break open the granulomas, induce an immune response, and target Mtb. A spray drying yield of 36.9% was obtained. In addition, the size of the SDP was 2.47 Â± 0.05 Î¼m, which is within the appropriate size range of 1-3 Î¼m for deep lung delivery. Our data shows that the BCG and INH loaded NPs, the combination of immuno- and chemotherapy, can successfully be incorporated in an inhalable dry powder and used as a potential delivery system in LTB patients. Further studies are needed to evaluate the immunogenicity of BCG and the toxicity of INH NPs in animal models. This inhalable dry powder may serve as the next step in controlling and eradicating LTB infections caused by one of the most virulent bacterial pathogens.

Controlled Drug Release with Ultrasound Imaging Monitoring Capability

Conference: Translational Medicine and Bioengineering Conference

Year: 2016

Proceeding: Translational Medicine and Bioengineering Conference

Group: Poster Submissions

Session: Poster Session

Time: Saturday, November 12, 2016 - 5:00pm-7:00pm

Authors:

Park, Y., University of Cincinnati

Over the past decade tremendous progress has been made in developing processes to direct the differentiation of human pluripotent stem cells to various specialized cell types for a variety of applications including development and disease modeling, drug screening and evaluation, and regenerative therapeutics. Typically these approaches involve construction of biomimetic microenvironments to guide stem cell differentiation in a developmentally-relevant manner. However, these approaches are often costly, inefficient, difficult to scale, and not compatible with cGMP manufacturing. The next generation of differentiation platforms requires simple, robust, inexpensive, and xeno-free characteristics. In this presentation I will discuss evolution of differentiation platforms for cardiovascular. We have identified canonical Wnt signaling as a key regulator of hPSC differentiation to cardiomyocyte and designed a protocol that produces high purity cardiomyocytes in a defined, xeno-free, growth factor-free system via appropriate temporal presentation of small molecule modulators of Wnt signaling. Furthermore, we have determined that a different canonical Wnt signaling profiles can direct hPSCs to vascular endothelial progenitors and epicardial cells in a defined, growth factor-free differentiation process. These epicardial cells can give rise to cardiac fibroblasts and smooth muscle cells upon induction of an epithelial-to-mesenchymal transition. Thus, by stage-specific modulation of canonical Wnt signaling and other developmental pathways we can generate a diverse array hPSC-derived cardiac cells in defined, xeno-free conditions for in vitro studies, drug screening and toxicology analyses, and development of regenerative therapies.

Detection of Tuberculosis Volatile Organic Biomarkers Methyl Nicotinate and Methyl Phenyl Anisate Using Two-Step Anodized Titanium Dioxide Nanotube Arrays

Conference: Translational Medicine and Bioengineering Conference

Year: 2016

Proceeding: Translational Medicine and Bioengineering Conference

Group: Poster Submissions

Session: Poster Session

Time: Saturday, November 12, 2016 - 5:00pm-7:00pm

Authors:

Saffary, Y., University of Utah
Willis, C., University of Utah
Misra, M., University of Utah
Mohanty, S., University of Utah

Rapid screening of tuberculosis (TB) by evaluation of associated volatile organic compounds (VOCs) in breath has been shown to be a promising technology that is significantly faster and more convenient than traditional sputum culture tests. Methyl nicotinate and methyl p-anisate have been isolated as specific biomarkers for the mycobacterium tuberculosis, as they are not found in high concentrations in ambient air or the breath of healthy patients, but should be readily present in all patients with active TB. These VOCs can be detected at very low levels using a solid-state sensor based on metal-functionalized 3D titanium dioxide (TiO_2) nanotube arrays that bind the biomarkers at specifically applied voltages. The change in current when the sensor is exposed to the target VOC is caused by binding to the TiO_2 nanotubes. The characteristic large resistance of the nanotube arrays on the titanium surface is extremely important in measuring these small (on the order of μA) current changes. Therefore, the structure and consistency of the TiO_2 is extremely important and methods for its improvement are constantly explored. The method described here is a two-step anodization, an improvement on the currently used one-step method. Previously, the nanotube layer has been grown in a single step on a polished titanium surface using a single voltage over a prescribed amount of time. It was found that anodizing at a higher voltage grew longer, more brittle tubes of larger diameter, the bulk of which can be easily removed using a simple adhesive, leaving only the strong base layer behind. This provides a consistent template on which the second layer of nanotubes can be deposited. The resulting arrays appear more highly organized when examined under SEM than their one-step method counterparts, displaying a honeycomb effect with fused sidewalls. Mimic testing in the lab can be used to test a sensors response to a given biomarker at known concentration, in this case sensors functionalized with cobalt chloride allow binding of methyl nicotinate at a specific voltage. After testing, it was found that sensors anodized using the two-step method not only provided more consistent current readings over the entire face of the sensor, but also showed faster response to the biomarker. Consistency over the surface is especially important in point of care TB sensors. In addition, the stronger more organized arrays allow for a reduction in noise when measuring current.

Elastin-like Polypeptides for Generation of Induced Pluripotent Stem Cells

Conference: Translational Medicine and Bioengineering Conference

Year: 2016

Proceeding: Translational Medicine and Bioengineering Conference

Group: Poster Submissions

Session: Poster Session

Time: Saturday, November 12, 2016 - 5:00pm-7:00pm

Authors:

Lee, C. H., Texas tech University
Gill, H., Texas Tech University

Generation of induced pluripotent stem cells using elastin like polypeptide
Chang hyun Lee[1], Harvinder S. Gill[1]

[1] Texas Tech University, Lubbock, Texas 79409, USA

Induced pluripotent stem (iPS) cells have been generated from various somatic cells using different approaches; however, a major restriction of reprogramming methods is the use of potentially harmful genome-integrating DNAs. Here, without a viral vector, we generated iPS cells from fibroblasts using an elastin- like polypeptide (ELP)-based transfection method. Our findings support the possible use of ELPs for transfer of the reprogramming genes in to somatic cells for generation of iPS cells. Results of dynamic light scattering (DLS) analysis demonstrated efficient complexation of ELPs with iPS genes. After transfection, the iPS cells showed ES cell-like characteristics, including expression of endogenous pluripotency genes, differentiation into three germ layer lineages, and formation of teratomas. Our results demonstrate that ELP-based gene delivery may provide a safe method for use in generation of virus-free and exogenous DNA-free iPS cells, which will be crucial for future applications in cell replacement therapy.

Keywords: IPS cells, Elastin like polypeptide,

Enhance Methanol Tolerability of M. Extorquens AM1 By Adaptive Evolution

Conference: Translational Medicine and Bioengineering Conference

Year: 2016

Proceeding: Translational Medicine and Bioengineering Conference

Group: Poster Submissions

Session: Poster Session

Time: Saturday, November 12, 2016 - 5:00pm-7:00pm

Authors:

Cui, L. Y., Tsinghua University
Liang, W. F., Tsinghua University
Zhang, C., Tsinghua University
Xing, X. H., Tsinghua University
Zhu, W. L., Tsinghua University

Chinese Hamster Ovary (CHO) cells are routinely optimized to stably express monoclonal antibodies (mAbs) at high titers. However, CHO cell technology can be unwieldy for rapid, high-throughput production and screening of candidate therapeutic proteins. Here, we have developed a cell-free protein biomanufacturing platform utilizing a commercially available CHO extract for the cell-free synthesis of mAbs. Our specific aims were to: (1) optimize cell-free reaction parameters to maximize protein yields, (2) screen additives to enhance disulfide bond formation for intact mAb production, and (3) establish the extract as a predictable tool for ranking yields of candidate antibodies. Optimization of the reaction conditions increased protein yields 8-fold to 700 mg/L using green fluorescent protein (GFP) as a reporter. Additives such as oxidized glutathione established an oxidizing cell-free environment for disulfide bond formation, and chaperones with disulfide isomerase activity further increased intact mAb yields 2-fold. Lastly, light and heavy chain expression was optimized to minimize aggregation products. Using our optimized platform, we demonstrate for the first time the cell-free synthesis of active, intact mAb using CHO extract. Additionally, the cell-free system demonstrated the capacity to rank known high and low producing mAb sequences in the same order as seen from CHO cell-based platforms. Unlike stable or transient transfection-based screening, which requires 8-14 days for setup and execution, results using our CHO-based CFPS system are attained within 2 days. Further development of the customizable cell-free platform would provide a tool for rapid, high-throughput ranking of mAb producers and has implications for the synthesis of difficult-to-express proteins.

Functional Ovarian Tissue Construct with Engineered Multi-Layered Follicles That Secrete Sex Hormones

Conference: Translational Medicine and Bioengineering Conference

Year: 2016

Proceeding: Translational Medicine and Bioengineering Conference

Group: Poster Submissions

Session: Poster Session

Time: Saturday, November 12, 2016 - 5:00pm-7:00pm

Authors:

Leung, R., Wake Forest Institute for Regenerative Medicine

Background: Ovarian hormone-deprived conditions result in a host of physiological conditions ranging from osteoporosis to urogenital disease. Hormone replacement therapy (HRT) is often prescribed to alleviate ovarian hormone cessation; however, pharmacological delivery methods consistently lead to increased serum concentrations and clinical complications such as cancer and heart disease.The present study aimed to overcome these drawbacks by delivering sex hormones by means of a functional ovarian tissue construct containing engineered ovarian follicles. The cell-based platform would permit interaction with the innate hypothalamic-pituitary-ovarian axis, allowing the regulated delivery of sex hormones at more physiologically-relevant levels.

Methods: Granulosa and theca cells were isolated from ovaries of 21-day old rats and used to generate three different types of ovarian tissue constructs. In the first scheme, both cell types were mixed together at a ratio of 3:1 granulosa cells to theca cells and subsequently encapsulated in a collagen cell carrier (referred to as co-cultured non-spheroid constructs). In the second scheme, prior to encapsulation, the mixed cells were formed into 3D spheroids through forced microwell aggregation (referred to as non-layered spheroid constructs). In the third scheme, a compartmentalized spheroid design was used (referred to as multi-layered spheroid constructs). First, spheroids composed solely of granulosa cells were formed. Theca cells were than aggregated over these spheroids, resulting in compartmentalized spheroids. To assess compartmentalization, spheroids were stained for FSHR (granulosa cell marker) and LHR (theca cell marker). The viability of encapsulated cells was determined using a live/dead assay. To evaluate the endocrine function of tissue constructs, the levels of 17 Î^{2}-estradiol (E2) and progesterone (P4) were measured in the culture media using ELISA kits.

Results: Immunofluorescent staining of the cellular spheroids revealed that microwell aggregation is capable of generating distinctly compartmentalized spheroids with structural architecture similar to that of native ovarian follicles. Live/dead staining of tissue constructs indicated that the viability of cells in 3D spheroids encapsulated in collagen are maintained over the course of 30 days in culture. Measurements of the amount of sex hormones in the culture media showed that non-layered spheroid constructs secrete significantly higher levels of E2 and P4 than co-cultured non-spheroid constructs ($p < 0.05$), and that multi-layered spheroid constructs secrete significantly higher levels of E2 and P4 than non-layered spheroid constructs ($p < 0.05$).

Conclusion: Our results demonstrate the feasibility of using a microwell-aggregation system to generate large numbers of homogenous follicle-like structures from primary endocrine cells, and the ability to leverage these engineered follicles to fabricate functional ovarian tissue constructs capable of sustained release of vital sex hormones. Furthermore, multi-layered spheroids that recapitulate the structural architecture of native ovarian follicles emphasize the importance spatial arrangement plays in maximizing the viability and function of ovarian endocrine cells. This study thus offers a simple but potent strategy for facilitating cell-based HRT and underlines the integral role tissue engineering strategies play in optimizing systems for cell therapeutics.

Improved Survival of Pancreatic Islets through Delivery of Alanine from Pegdma Microparticles

Conference: Translational Medicine and Bioengineering Conference

Year: 2016

Proceeding: Translational Medicine and Bioengineering Conference

Group: Poster Submissions

Session: Poster Session

Time: Saturday, November 12, 2016 - 5:00pm-7:00pm

Authors:

Long, A., University of California, Berkeley
Le, L.
Weber, R.
Faleo, G.
Desai, T., University of California, San Francisco

Type 1 diabetes is an autoimmune disease characterized by the selective destruction of insulin producing Î²-cells within the pancreas. The disease has no established cure, with typical treatments ranging from insulin therapy to whole pancreas and islet transplantation. Cell replacement therapy through transplantation of islets or stem cell-derived Î²-cells can facilitate complete insulin independence, and is thus a promising treatment for type 1 diabetes. However, protecting these cells from the bodyâs inflammatory response represents a formidable challenge to the success of the transplantation process. Recent studies have shown that amino acid supplementation promotes islet survival in high-density cultures that mimic the hostile transplant environment. Thus, a biocompatible, injectable nutrient delivery system may hold the key to in vivo islet longevity during the post-transplantation period. In this work, we present two therapies engineered to satisfy the role of nutrient supplementation in islet survival. First, photolithography was used to generate poly(ethylene glycol) dimethacrylate (PEGDMA) hydrogel microparticles that are loaded with alanine. The release profile of alanine from these microstructures was optimized by changing microparticle cross-linking densities and amino acid concentrations. Our data suggest that 20% PEGDMA microparticles with a 600 mM loading concentration of alanine are effective in preventing islet death. Second, PEGDMAÂ microparticles and beta cells were functionalized with complementary DNA strands to promote cell adhesion and nutrient delivery. Efficient targeting of fatty-acid modified oligonucleotides to Î2-cell membranes was achieved through stepwise assembly, and cell-particle adhesion was examined. Both of these therapies satisfy the role of nutrient supplementation in islet survival and can enhance the therapeutic benefits of islet transplantation as a treatment for type 1 diabetes.

Inquiry into Neurodegeneration: Engrailed Proteins and Parkinson's Disease

Conference: Translational Medicine and Bioengineering Conference

Year: 2016

Proceeding: Translational Medicine and Bioengineering Conference

Group: Poster Submissions

Session: Poster Session

Time: Saturday, November 12, 2016 - 5:00pm-7:00pm

Authors:

Koziol, H., Montana State University

Parkinsonâs disease is a devastating neurodegenerative disease that is characterized by the depletion of dopaminergic neurons in the midbrain, leading to a decreased supply of dopamine and involuntary muscle control. An estimated 10 million people worldwide have Parkinsonâs disease and approximately 60,000 Americans are newly diagnosed with this disease each year. The United States, like other developed nations, has a growing need for interventions for Parkinsonâs disease. The number of patients with Parkinsonâs disease (PD) is expected to rise as the average age of the US population is increasing.

In spite of the frequency of this disease, there is little understanding of the cause of Parkinsonâs disease, though researchers have reported that genetics and the environment may be significant factors in the etiology of the disease. Studying this disease in vivo is difficult because it would require access to the human midbrain in diseased patients, a feat which at the time is not feasible. Therefore, in vitro techniques such as the use of stem-cell based models can be used to study the disease and find potential treatments. Stem cells derived from the reprogrammed skin cells of PD and non-PD patients can be derived and used to grow dopaminergic neurons in vitro. Previous research has suggested that dopaminergic neurons derived from patients with PD are more sensitive to cytotoxic insult than non-PD derived dopaminergic neurons. While the exact cause of the death of these neurons in patients is unknown, it is postulated that the aggregation of alpha-synuclein (SNCA) is one of the primary culprits. Our research group recently gathered evidence that Engrailed proteins and alpha-synuclein interact physically. The Engrailed homeoproteins (Engrailed-1 (En-1)/ Engrailed-2 (En-2)) are believed to promote differentiation of mesencephalic dopaminergic (mDA) neurons during development and to have a neuroprotective effect throughout the lifetime of the mDA neuron. These proteins have been shown to protect mDA neurons in Parkinsonâs disease animal models and enhance dopamine synthesis.

My hypothesis is that Parkinsonâs disease may be in part due to high levels of alpha-synuclein having an inhibitory effect on Engrailed proteins. I am exploring the hypothesis to investigate whether Engrailed proteins protect neurons from cell death, which are linked to high levels of alpha-synuclein. To do this, I am creating six genotypes from dopaminergic progenitor cells to test genes and genetic interactions between the two Engrailed proteins and alpha-synuclein. I will then use traditional mDA differentiation models to assess the effects of SNCA and EN proteins on dopaminergic neuron survival, as assayed using a TH-luciferase reporter gene to measure DA-neuron cell number. I expect an increase in dopamine activity from the doubly overexpressed alpha-synuclein and Engrailed protein cell cultures because their overexpressed effects should counteract one another compared to alpha-synuclein overexpression alone. By exploring these combinations of genotypes, I hope to shed light on the genetics behind Parkinsonâs disease, and how increased levels of alpha-synuclein may lead to the death of midbrain dopaminergic neurons.

Learning How Immune Cells of the Gut Contribute to Autoimmune Disease through Reverse-Engineered Transcriptional Regulatory Networks

Conference: Translational Medicine and Bioengineering Conference

Year: 2016

Proceeding: Translational Medicine and Bioengineering Conference

Group: Poster Submissions

Session: Poster Session

Time: Saturday, November 12, 2016 - 5:00pm-7:00pm

Authors:

Miraldi, E. R., Simons Foundation // NYU School of Medicine
Pokrovskii, M.
Hall, J. A.
Martins de Castro, D., New York Univeristy
Carriero, N.
Watters, A.
Littman, D. R.
Bonneau, R., New York University

The National Institute of Health estimates that at least 23.5 million Americans suffer from autoimmune disease, yet even the most prevalent autoimmune diseases have no cure, requiring costly lifelong immuno-suppressive therapies. Genome-wide association studies (GWAS) provide lists of genes associated with autoimmune diseases, and many of these genes are specifically expressed by diverse immune cell populations, including arms of both the adaptive and innate immune system. Many of these genes are involved in immune-cell sensing and response to the microbiota. As the gut contains the highest concentration of host-associated microbiota, we reasoned that characterization of gene expression by RNA-seq in gut-associated immune cell populations would implicate specific populations in disease etiology. In this study, we focus on innate lymphoid cells (ILCs) under homeostatic conditions and in response to microbial perturbations in mice. Furthermore, we wanted to design perturbations (e.g., gene knock down) for specific populations to change phenotypes that might contribute to autoimmunity. Thus, we not only set out to characterize transcriptional responses but also to infer mechanisms of transcriptional control, to describe gene expression as multivariate functions of transcription factor (TF) activities. TF gene expression measurements are often a poor proxy for protein TF activities, so we integrated chromatin accessibility (ATAC-seq) measurements to get noisy sets of signed, TF-gene interactions and then use penalized matrix decomposition methods to estimate TF activities and reverse-engineer transcriptional regulatory networks. We connected specific TFs (on the basis of predicted targets), in specific ILC populations, with autoimmune and other diseases. Predicted TF-gene regulatory hypotheses are currently being tested experimentally.

Mesenchymal STEM CELLS As Trojan Horses for the Selective Delivery of Nanoparticles to Tumors

Conference: Translational Medicine and Bioengineering Conference

Year: 2016

Proceeding: Translational Medicine and Bioengineering Conference

Group: Poster Submissions

Session: Poster Session

Time: Saturday, November 12, 2016 - 5:00pm-7:00pm

Authors:

Santamaria, J., University of Zaragosa
Encabo-Berzosa, M. M., Universidad de Zaragoza
Arruebo, M., University of Zaragosa
Gimeno, M.
Sancho-Albero, M.
Lujan, L.
Martin-Duque, M. P.
Gomez, L.
Quintanilla, M.
Sebastian, V., University of Zaragosa

A continuing challenge of cancer therapy is to attain sufficient selectivity by focusing the treatment in tumor cells, to avoid or minimize systemic side effects. Mesenchymal stem cells (MSCs) can be used as therapeutic vectors because of their ability to migrate and incorporate into inflammation areas such as tumors. Here, this homing ability is exploited to carry therapeutic nanoparticles (hollow Gold Nanoparticles (HGNs)) by a âTrojan-horseâ strategy. Â Amongst the different nanoparticles that could be employed, HGNs have the capacity to resonate in the near infrared region when irradiated by a suitable light source. The absorbed energy is transformed into heat causing local hyperthermia and killing nearby tumor cells. At these wavelengths healthy tissues have a minimal light absorption, so the effect is restricted to the tissues containing HGNs. However, delivering a dose of HGNs to the tumor that is sufficient to achieve the desired thermal efects is challenging when the HGNs are intravenously administered, in spite of the so-called EPR effect that helps to accumulate nanoparticles of suitable size in tumors. This was clear in the results of our experiments with U251MG cell-line xenograft tumors, where a poor therapeutic result was obtained for PEG-HGNs injected intravenously in spite of their PEG coating. Two factors are likely contributing to this effect, on the one hand, the intrinsic heterogeneity of the EPR effect that leads to a variable distribution of plasmonic nanoparticles within the tumor. On the other, the fact that a significant fraction of the free PEG-HGNs injected intravenously are detected and cleared by RES macrophages, as shown by the gold accumulation found in the spleen and liver of the test animals, reducing the delivery of HGNs to the tumor Both problems were alleviated by using MSCs as carriers of those nanomaterials.

On the other hand, a successful internalization of both bare and PEGylated hollow gold nanoparticles was demonstrated in murine MSCs incubated with HGNs under suitable conditions. The HGNs remain inside the cells for at least one week, a period sufficient for the MSCs to reach tumor sites. Taking advantage of the tumor-homing ability of nanoparticle-loaded MSCs, a more homogeneous nanoparticle distribution across the tumoral tissue was obtained, as demonstrated by histological analysis of treated tumors, and this phenomenon increased the efficiency obtained when applying photothermal therapy on subcutaneous xenograft glioma. The immunohistochemical analysis also confirmed the efficiency of photothermal theraphy, especially in those tumors where gold nanoparticles had been delivered by MSCs. Therefore the Trojan Horse strategy outperforms the EPR (Enhanced Permeation and Retention) effect, commonly exploited in the delivery of nanoparticles to tumors.

Finally, in this work we also demonstrate how MSCs internalize the HGNs and after intravenous administration they reach the tumors still loaded with HGNs. Interestingly, internalization of HGNs affects the motility of MSCs (as evidenced by would healing experiments) and the expression of some key constituent proteins. This opens up numerous opportunities to exploit MSCs as carriers of different functional nanoparticles.

N-Acetylcysteine-Doped Microparticles Efficiently Rescue Oligodendrocyte Precursor Cell Redox State in an Oxidative Environment

Conference: Translational Medicine and Bioengineering Conference

Year: 2016

Proceeding: Translational Medicine and Bioengineering Conference

Group: Poster Submissions

Session: Poster Session

Time: Saturday, November 12, 2016 - 5:00pm-7:00pm

Authors:

Lampe, K., University of Virginia
Murphy, N.

Introduction: Overproduction of reactive oxygen species (ROS) is a major factor in the pathogenesis of multiple sclerosis[1]. Localized antioxidant delivery is therapeutically relevant to counteract and neutralize ROS in such oxidant-stressed systems. Poly(lactic-co-glycolic acid) (PLGA) nanoparticles can provide a localized, controlled release of encapsulant, though reported encapsulation efficiencies of small-molecule therapeutics have been low: 12%[2] and 16%[3] for dopamine and N-acetylcysteine (NAC), respectively. Here we investigate how modifying the traditional double-emulsion, solvent evaporation method by doping external water phases with the small-molecule drug affects encapsulation efficiency, temporal release profiles, released drug activity, and released drug protective effect on oligodendrocyte progenitor cell (OPC) viability after H_2O_2 stress. We aim to deliver NAC, a potent, cheap, FDA-approved antioxidant, with high efficiency and temporal control.
Â

Materials and Methods: NAC-loaded PLGA nanoparticles with varying lactide:glycolide (L:G) ratios and inherent viscosities were prepared by the double-emulsion, solvent evaporation method. The solvent for all water phases was deionized water. Briefly, the internal water phase (10% w/v NAC) was dispersed via sonication within the oil phase (15% PLGA in dichloromethane). The primary emulsion was dispersed in the outer water phase (1% PVA with or without 2.5% NAC) via sonication. The secondary emulsion was stirred in 0.3% PVA with or without 5% NAC for 4 hours. Particles were washed 3x in either DI water or 2.5% NAC and lyophilized. Release was determined by incubating particles at 37Â°C in phosphate-buffered saline and measuring supernatant NAC via Ellmanâs Reagent. LAP photoinitiator radical concentrations were measured via luminol to determine the dose response of NAC radical scavenging. OPC viability in response to oxidative conditions with and without the NAC antioxidant was measured via the CellGlo reagent.

Â

Results and Discussion: Encapsulation efficiencies (relative to theoretical load without doped external phases) for NAC-loaded PLGA nanoparticles were greatest with external phase doping â reading greater than 300% of the case when undoped. With the doped particles, NAC was released for 14 days. NAC ranging from 0.5-1000 ÂµM (the particle supernatant concentration range for each sample time) scavenged photoinitiated LAP radicals in a dose-dependent manner, with 0.5 ÂµM and 1000 ÂµM NAC scavenging 6.82% and 95.5% of radicals, respectively. Ongoing experiments show NAC-loaded nanoparticle antioxidant activity and protective effect on OPC viability.

Â

Conclusions: Drug-doping of external water phases during the double emulsion, solvent evaporation method gave NAC-loaded PLGA nanoparticles with significantly higher encapsulation efficiencies than without doping, and up to 20-fold higher than previously reported values. NAC scavenged LAP radicals in a dose-dependent manner and was released from PLGA nanoparticles for at least two weeks. This indicates that these particles will provide a larger and more continuous small molecule dose longer than previously reported methods. Efficiently loaded antioxidant nanoparticles capable of controlled and localized delivery could have therapeutic potential in oxidant-stressed systems and bias progenitor cells such as OPCs toward a self-renewing fate.

References: [1] Lehnardt S. J. Neurosci, 2002. 22:2478-86. [2] Pahuja R. ACS Nano, 2015. 9:4850-71. [3] Lancheros R. Univ. Sci., 2014. 19:161-168

Nanoparticle Therapeutics for Treating Cancer and Other Diseases of the Brain

Conference: Translational Medicine and Bioengineering Conference

Year: 2016

Proceeding: Translational Medicine and Bioengineering Conference

Group: Poster Submissions

Session: Poster Session

Time: Saturday, November 12, 2016 - 5:00pm-7:00pm

Authors:

Davis, M. E., California Institute of Technology

We have translated three nanoparticles into the clinic for use as cancer therapeutics. One of these nanoparticle, experimental therapeutics was the first to show functional RNA interference (RNAi) in humans. At this time, these nanoparticles have been investigated in over 10 different clinical trials. I will describe how we translated these nanoparticles from the laboratory at Caltech into clinical, experimental therapeutics. In addition to investigations of nanoparticle-based therapeutics for cancer, we are now concentrating on developing nanoparticles that cross the blood-brain-barrier for treating cancer and other diseases in the brain. Our recent efforts will be presented.

Next-Generation Sequencing and Omics Technologies for Biotherapeutic Manufacturing Development

Conference: Translational Medicine and Bioengineering Conference

Year: 2016

Proceeding: Translational Medicine and Bioengineering Conference

Group: Poster Submissions

Session: Poster Session

Time: Saturday, November 12, 2016 - 5:00pm-7:00pm

Driven by a continuous innovation of experimental and computational methodologies, Next-Generation Sequencing (NGS) and other Omics technologies continue to advance at a rapid pace. These technologies have led to a revolution in biological research and medicine but have applications well beyond those areas. One field in which these technologies have started to make a positive impact is the development of cell lines and bioprocesses for biotherapeutic manufacturing.

At Biogen, we are using NGS to both characterize therapeutic production clones and evaluate manufacturing culture processes, many of which use Chines Hamster Ovary (CHO) cells. One application is an NGS screen of candidate production clones to ensure their biotherapeutic mRNA sequence is free of mutations. For this screen we developed a targeted mRNA sequencing assay that multiplexes samples of multiple clones onto a single sequencer lane. This mutation screen has become a standard part of our cell line development workflow to which it adds quality and rigor on decisions about candidate production clones. Another application uses targeted genome sequencing to characterize integration sites of the biotherapeutic transgene. This integration site assay allows for clone fingerprinting, provides information on genetic stability, and supports efforts for locus specific engineering of transgene integration into the CHO genome.

In addition to clone characterization we use NGS (RNA-Seq) to obtain insights into the cell biology of bioprocessing. The genome for CHO is much less resolved than the genome for human or mouse and to set up a robust omics platform around CHO we evaluated several draft CHO genome assemblies and annotations. In addition, we developed a standardized computational pipeline to process CHO RNA-Seq data. This enabled comparing gene expression profiles across our bioprocess which led to the identification of changes related to particular growth stages and transgene expression.

In conclusion we successfully developed and implemented a toolbox of NGS technologies that we continue to refine and expand. These NGS tools provide improvements to our cell line development workflows and generate new insights into our CHO production processes.

On-Demand Nanovaccines Manufacturing

Conference: Translational Medicine and Bioengineering Conference

Year: 2016

Proceeding: Translational Medicine and Bioengineering Conference

Group: Poster Submissions

Session: Poster Session

Time: Saturday, November 12, 2016 - 5:00pm-7:00pm

Authors:

Crecente-Campo, J., University of Santiago de Compostela

Our laboratory has developed technologies for the production of polymer nanocapsules consisting of different combinations of oils and polymers, notable chitosan [1], hyaluronic acid [2], and polyglutamic acid [3]. These nanocapsules have been generally produced by the solvent displacement technique, which so far has allowed us the formation of nanocapsules with a size in the range of 100-300 nm.
With this work we aimed to achieve a double goal: from a technological perspective, our objective has been to identify the critical parameters that influence the size and the polymer deposition (one or multiple polymer layers) around the oily droplets; and, in that way, to obtain nanocapsules of well defined sizes and with different surface structures. Besides, from a biomedical point of view, the goal was to study the potential of selected nanocapsules as nanocarriers for vaccines antigens.

Two achieve these two goals, NCs of chitosan/lecithin/vitamin E have been used as a model system. The ratio between organic and aqueous phase, as well as the pressure of the addition of the first one over the second one, have shown to be the most critical parameters that affect the final size of nanocapsules (from 50 nm up to 500 nm). Moreover, nanocapsules with different layers of oppositely charged polymers in the shell could be easily prepared by ionic interactions. With this mild method we built up to 5 layers of chitosan-dextran sulfate surrounding the lecithin/vitamin E core.

Finally, to show the potential of these nanosystems as carriers for vaccination, a model antigen from E. coli was associated to the first layer of both, the monolayer and the bilayer nanocapsules. The association efficiency was high in both cases (67% and 71%). The second layer of dextran sulfate makes the formulation more stable after freeze-drying and retards the release of the antigen. In vitro studies in macrophages have shown moderate toxicity, good cell internalization and the production of cytokines related with the activation of the macrophages. With all these data we envisage that both monolayer and bilayer nanocapsules are promising candidates to test in vivoas thermostable vaccines.

[1] S. Vicente, B. Diaz-Freitas, M. Peleteiro, A. Sanchez, D.W. Pascual, A. Gonzalez-Fernandez, et al., A Polymer/Oil Based Nanovaccine as a Single-Dose Immunization Approach, PLoS One. 8 (2013) 2â9. doi:10.1371/journal.pone.0062500.

[2] F.A. Oyarzun-Ampuero, G.R. Rivera-Rodriguez, M.J. Alonso, D. Torres, Hyaluronan nanocapsules as a new vehicle for intracellular drug delivery, Eur. J. Pharm. Sci. 49 (2013) 483â490. doi:10.1016/j.ejps.2013.05.008.

[3] E. Borrajo, R. Abellan-Pose, A. Soto, M. Garcia-Fuentes, N. Csaba, M.J. Alonso, et al., Docetaxel-loaded polyglutamic acid-PEG nanocapsules for the treatment of metastatic cancer, J. Control. Release. 238 (2016) 263â271. doi:10.1016/j.jconrel.2016.07.048.

Point of Care Screening of Tuberculosis from Breath Using Functionalized TiO2 Nanotube Array Sensing Platform

Conference: Translational Medicine and Bioengineering Conference

Year: 2016

Proceeding: Translational Medicine and Bioengineering Conference

Group: Poster Submissions

Session: Poster Session

Time: Saturday, November 12, 2016 - 5:00pm-7:00pm

Authors:

Mohanty, S., University of Utah
Saffary, Y., University of Utah
Willis, C., University of Utah
Misra, M., University of Utah

Improving screening and diagnosis of tuberculosis (TB) is critical to reducing its burden around the world. Some of the key challenges include that up to 1/3rd of adults and most young children with TB do not produce sputum and, even among those who do, bacillary burden in sputum is often low. Due this these reasons, the World Health-Organization has ranked rapid non-sputum based biomarker based tests has the highest priority for diagnostics and screening of TB. One approach is to evaluate volatile organic compounds (VOCs) in exhaled breath, which is easily collected from patients of all ages. Recent research has identified at least two VOCs â methyl nicotinate and methyl p-anisate â that are a by-product of Mycobacterium tuberculosis metabolism and should therefore be present in anyone with active TB, regardless of age. These VOCs do not appear to be emitted by other bacteria of the respiratory tract; do not appear to be found at high concentrations in the ambient environment; and have been identified at high concentrations in the breath of TB patients but not healthy controls. However, the lack of a technology to easily measure specific VOCs at the point-of-care has been a key barrier to further validation and use of these VOCs as a biomarker for TB diagnosis.
This work will describe a low-cost, handheld device for identifying target VOCs in exhaled breath. The device is a solid-state sensor based on metal-functionalized 3D titanium dioxide (TiO_2) nanotube arrays that bind specific VOCs when a specific voltage gradient is applied. Binding results in a change in current that is proportional to the concentration of target VOCs

present in the breath. The sensor has been customized to specifically bind methyl-nicotinate and methyl p-anisate and has a limit-of-detection that is several orders of magnitude below levels of the two VOCs observed to date in TB patients. Furthermore, the detection is fast (on the order of minutes), reagent-free, and requires no sample preparation. The concept of the point-of-care breath sensor device is shown in Figure 1a.Â Device is portable, operated using a smartphone, and can be taken directly to the patientâs home which is important for individuals that live far from medical facilities as is often the case in high TB burden countries.

Results of the sensor showed the TiO2 sensor gave a positive response to breath taken from TB positive patients and no response to breath taken from health individual controls.

Initial results of the TiO2 nanotube sensor shows a correlation between a high sensor response and TB positive patient. Site studies are currently underway in Uganda and India to establish the feasibility of using the technology in the field as a point of care screening/diagnostic device. A summary of these results will be presented.

Pollen Grains for Oral Vaccination

Conference: Translational Medicine and Bioengineering Conference

Year: 2016

Proceeding: Translational Medicine and Bioengineering Conference

Group: Poster Submissions

Session: Poster Session

Time: Saturday, November 12, 2016 - 5:00pm-7:00pm

Authors:

Gill, H. S., Texas Tech University

Oral vaccination continues to be one of the holy grails in vaccinology. Not only is oral vaccination needle-free, convenient, painless, amenable to self-administration, child-friendly, less expensive, safer because needle stick injuries are eliminated, but it can also induce mucosal immunity. Mucosal surfaces are a major portal of pathogen entry, and mucosal immunity has the potential to neutralize pathogens before they can cause infection. Oral vaccination is however difficult to achieve due to the acidic and harsh enzymatic environment of the stomach. We have recently developed pollen shells as a vehicle for oral vaccination. We did studies to evaluate (i) the ability of pollen shells from different plant species to induce a systemic and mucosal immune response in a mouse model using ovalbumin as an antigen, (ii) the durability of the immune response, and (iii) the cytokines secreted by mouse bone marrow derived dendritic cells (mBMDCs) and macrophages in the presence of pollen shells to better understand the mechanism of pollen shells for oral vaccination. Additionally, we also performed in vitro studies to evaluate cytotoxicity of pollen shells by incubating them with Caco-2 cells (to model human epithelial cells). Overall our studies show that pollen shells stimulate both systemic and mucosal immunity thus suggesting that they have good potential to be used for oral vaccination. The in vitro experiments show that pollen shells can activate macrophages and DCs, and thus provide critical information about the mechanism of the potential adjuvant property of pollen shells for oral vaccine delivery.

Production of Designer Lipids Using "Click" Chemistries

Conference: Translational Medicine and Bioengineering Conference

Year: 2016

Proceeding: Translational Medicine and Bioengineering Conference

Group: Poster Submissions

Session: Poster Session

Time: Saturday, November 12, 2016 - 5:00pm-7:00pm

Authors:

Konetski, D., University of Colorado
Bowman, C. N., University of Colorado
Wang, C., University of Colorado
Worrell, B., University of Colorado
Gong, T., University of Colorado
Zhang, X.
Baranek, A., University of Colorado
Xi, W., University of California, Los Angeles
Chatani, S.

Liposomes have proven to be excellent drug delivery vehicles and as such have generated great interest in current research. However, in order for their utility to be realized, liposomes frequently require chemical modifications to the phospholipid structure. To this end, we demonstrate a simplified method for the synthesis of designer lipids through the use of a functionalized lysolipid to allow for integration of selected lipid tails. Using a variety of functionalities, we have established âclickâ based phospholipid formation chemistries such as Copper-catalyzed Azide-Alkyne Cycloaddition (CuAAC), thiol-Michael, and thiol-acrylate initiated upon the introduction of a range of catalysts including reducing agents, base, or light. This range in formation chemistries and initiation conditions allows for the incorporation of useful functional groups into the phospholipids with the high yield and mild conditions characteristic of âclickâ reactions without the new functionalities becoming compromised. Following formation, we have demonstrated this orthogonality through the inclusion of crosslinkable groups for enhanced stability. In addition, our thiol functionalized lysolipids are capable of formation viatransthioesterification in a catalyst free environment, allowing for future lipid remodeling. Together, these new chemistries will open the door for simplified exploration of improved liposomal drug carriers without the need for labor-intensive chemical synthesis.

Proof-of-Concept Study for Rapid Detection of Zika Viral RNA Via DNA Biosensor

Conference: Translational Medicine and Bioengineering Conference

Year: 2016

Proceeding: Translational Medicine and Bioengineering Conference

Group: Poster Submissions

Session: Poster Session

Time: Saturday, November 12, 2016 - 5:00pm-7:00pm

Authors:

Dylewski, H. C., University of Tennessee Knoxville
Eda, S., University of Tennessee Knoxville
Ramos, J., University of Tennessee Knoxville
Rice, H. J., University of Tennessee Knoxville

Zika virus is a flavivirus that causes disease in humans. Zika infection has been linked to severe medical conditions such as Guillain-Barre syndrome in adults and microcephaly in newborns of mothers infected during pregnancy. The possibility of severe consequences has led to the need for new diagnostic techniques to replace the current inefficient and time consuming methods. We conducted a proof-of-concept study to demonstrate that a selective DNA-biosensor with peroxidase activity is able to specifically detect DNA molecules with a Zika virus-specific sequence (zDNA). Our approach utilizes a G-quadruplex hemin binding DNA aptamer, an effective signaling mechanism that forms a DNAzyme with peroxidase activity able to catalyze the oxidation of tetramethylbenzidine (TMB). The enzymatic activity of the DNAzyme, termed Proxzyme, is activated only in the presence of zDNA through a proximity dependent DNAzyme formation mechanism. In a series of experiments we optimized (1) reaction buffer composition, (2) length of spacer sequence, and (3) DNAzyme type. The developed zProxyzyme is proven to selectively detect zDNA in the nano-molar range in under 16 minutes colormentrically. The biosensor shows promising selectivity and will be paired with an on-site diagnostic device previously developed by our lab that utilizes alternating current electrokinetic (ACEK) based capacitance sensing to decrease the limit of detection.

Self-Aggregation of Functionalized Capsids Using Coarse Grained Molecular Dynamics

Conference: Translational Medicine and Bioengineering Conference

Year: 2016

Proceeding: Translational Medicine and Bioengineering Conference

Group: Poster Submissions

Session: Poster Session

Time: Saturday, November 12, 2016 - 5:00pm-7:00pm

Authors:

Karra, V., Rutgers University
Libring, S., Rutgers University
Chong, L., NETL
Dutt, M., Rutgers University

Viral capsids, the protein shells of a virus, have been the subject of significant amount of research in a variety of fields. Their potential relevance to biomedical science arises from their ability to encapsulate or carry drugs and deliver them to targeted sites in the body. The capsids can be decorated with polymers, long chains of hydrocarbons, to target specific sites or promote controlled aggregation for creating novel materials. Since viral capsids are within the same scale as nanoparticles, a cluster of them have a larger surface area, which is more area for the drugs to be attached to the capsids. We used coarse-grained molecular dynamics to model and simulate the aggregation of the Cow-Pea Mosaic Virus (CPMV) capsids decorated with polyethylene glycol (PEG), our polymer. Additional variations of the system includes changing the polymer density on the capsids and the capsid volume fraction, to show how these changes affect the self-assembled structures. We analyzed the systems to find the overall shape of the clusters, using the cluster size, the number of neighbors for each capsid, the effective radius of the cluster, and the moments of inertia. Our observations suggest that (a) the excluded volume of the capsids and (b) the polymer density on the capsids play the largest role on the aggregation dynamics when compared to other variants. Our findings can help further the research of targeted drug delivery systems using nanoparticles decorated with polymers.

Synthesis and Enantioselective Transformation of a New Series of Pharmacologically Important 4-Aryl-1, 4-Dihydropyridines By Microbial Lipases

Conference: Translational Medicine and Bioengineering Conference

Year: 2016

Proceeding: Translational Medicine and Bioengineering Conference

Group: Poster Submissions

Session: Poster Session

Time: Saturday, November 12, 2016 - 5:00pm-7:00pm

Authors:

Kaur, J., UIET, Panjab University
Kaur, A.

Dihydropyridines are the most important class of calcium antagonists having broad range of pharmaceutical and therapeutic effects. Second generation dihydropyridines possess at least one chiral centre and the effects differ from one enantiomer to the other. The other enantiomer of the drug is many a times an antagonist, leading to serious side effects upon administration. The present work describes the synthesis of new series of 4-aryl-1, 4-dihydropyridines. Microwave assisted synthesis of different dihydropyridine analogues have been studied in our laboratory, offering distinct advantages of shorter reaction time and avoiding hazardous solvents. The compounds so obtained were analysed via NMR, IR and MS for structure elucidation. These compounds were screened in terms of wavelength, temperature and eluent composition to resolve the components in the shortest time. These racemic analogues were then resolved on a Chiralcel ODH column of HPLC. The wavelength so chosen is based on the substantial absorption by the compound possessing the best S/N ratio. This was first assayed with a mobile phase comprising of n-hexane and isopropanol in different proportions between 22 to 30°C. The racemic analogues were resolved best at 353nm with a retention time of less than 10 minutes. Further the stereo selective biotransformations have been investigated on these racemic analogues using Candida antarctica lipase. The optical rotation values were measured using polarimeter. The non-zero values for the optical rotation confirms the enantioselective hydrolysis. The optical yields and optical configurations are being investigated by HPLC where an enantiomeric access of 65-70% has been attained.

The Protein Data Bank Japan (PDBj): Uncovering Hidden Trends in Macromolecular Structure Data

Conference: Translational Medicine and Bioengineering Conference

Year: 2016

Proceeding: Translational Medicine and Bioengineering Conference

Group: Poster Submissions

Session: Poster Session

Time: Saturday, November 12, 2016 - 5:00pm-7:00pm

Authors:

Ragsac, M., University of California, San Diego
Nakamura, H.
Kinjo, A.

The transfer of tumor-targeting T cells to patients is a promising approach for cancer immunotherapy. In particular, chimeric antigen receptors (CARs) modified T cells had demonstrated phenomenal efficacy against lymphoid leukemia, with more than 90% complete remission being observed. Despite these encouraging results, many challenges remain to be addressed before T cell therapy can be widely adopted for other cancers. Limiting the toxicity without comprising their efficacy are some of the key hurdles that need to be overcome. Here I will describe our recent efforts in using synthetic biology to develop drug-controllable genetic circuits and signaling switches for temporally modulating T cell activation. Specifically, I will highlight a set of multi-state memory switches for regulating multiple CARs expression, which enable the tuning of CAR expression level and ligand specificity. In addition, I will also discuss our novel dual small-molecule-gated signaling switches for ON/OFF control (one drug for ON switch, one drug for OFF switch) of T cell activation originated from T cell receptors and CARs. These genetic tools will provide a flexible tool kit for managing or preventing the toxicity associated with T cell therapy.

Theoretical and in Vitro Understanding of the Effect of Ligand-Receptor Valency on Dual-Ligand Nanoparticle Gene Delivery to Cancer Cells

Conference: Translational Medicine and Bioengineering Conference

Year: 2016

Proceeding: Translational Medicine and Bioengineering Conference

Group: Poster Submissions

Session: Poster Session

Time: Saturday, November 12, 2016 - 5:00pm-7:00pm

Authors:

Levine, R. M., University of Minnesota
Kokkoli, E., University of Minnesota

Cancer biomarker targeting has greatly improved cancer specific delivery of therapeutics, but its effectiveness is limited by on-target, off-tumor delivery. It was proposed that off-tumor delivery can be minimized with simultaneous targeting to two cancer specific receptors, requiring the high expression of both receptors uniquely present in cancer cells for efficient delivery. To test this hypothesis, we produced dual-ligand stealth liposomes functionalized with Î±$_5$Î²$_1$ and Î±$_6$Î²$_4$ integrin specific ligands. Binding affinity of multivalent nanoparticles is dependent on ligand valency, receptor valency and monovalent ligand-receptor affinity. Therefore, heteromultivalent nanoparticle binding is a complex interaction dependent upon valency and affinity of each ligand-receptor pair, as well as the ratios of these variables. A kinetic model was created to guide the design of this complex system. Dual-ligand liposomes were produced with varying ligand ratios and delivered to model cell lines with varying expression levels and ratios of the Î±$_5$Î²$_1$ and Î±$_6$Î²$_4$ integrins. Nanoparticle binding and internalization as well as integrin internalization rates were also evaluated to understand the effect of valency and avidity on delivery. Results showed that of all formulations and cells tested, dual ligand liposomes with equal ligand valencies achieved enhanced binding and selectivity for cancer cells expressing equal and high levels of receptor expression. These trends were consistent between theoretical and experimental results. The optimized liposomes were further used to achieve efficient and selective transfection in dual expressing cancer cells. With a quantitative understanding of dual ligand liposome binding, the insights gained from this study can inform future design of heteromultivalent nanoparticles for cancer therapy delivery.

Towards the Development of Robust and Meaningful Potency Assays for MSCs

Conference: Translational Medicine and Bioengineering Conference

Year: 2016

Proceeding: Translational Medicine and Bioengineering Conference

Group: Poster Submissions

Session: Poster Session

Time: Saturday, November 12, 2016 - 5:00pm-7:00pm

Authors:

Abukar, F.

In the UK alone, there are 146,000 people that have a Myocardial Infarction (MI) every year. Multiple clinical trials have explored MSCs for therapeutic application for ischemic cardiac injury. Currently there are no potency assays in the clinical setting that give a reliable measurement of MSCs mode of action, other than simple surface marker expression, which does not predict function. For this reason, there is a need to develop in vitro assays that will be predictive of functional activity of the MSC product. To be predictive, the assays need to be performed using parameters that more closely mimic the environment in the body, because measuring cell attributes in ambient conditions does not inform on how the cells will behave in vivo. Vascular assays can be used as platforms to study the hypothesis that MSCs functionally support blood vessels as a key mode of action in ischemic injured tissues. Vascular assays have traditionally lacked robustness and reproducibility, particularly since biological materials are highly variable. Our work focuses on developing robust assays with increased reproducibility. To achieve this, one aim of the project is to use Design of Experiments (DOE) to systematically optimize key parameters that affect the tubule-formation capacity of human endothelial cells, with an aim of making the assays more standardized and lowering assay costs in the process. In the current work, we found that tubule network formation could be controlled using lower quantities of an optimized matrix hence increasing the number of assays achievable from a single batch of substrate. Previous literature has indicated that efficiency of tubule forming capacity decreases drastically after endothelial cells reach passage 10. Therefore further work was conducted to investigate the impact of passage number on tubule formation capacity on the optimized matrix. Automated cell counting methods were used and were validated using manual counts. Here we show that higher passages than the ones presented in the literature, are still able to form networks when using optimized matrices, therefore increasing the possibility for higher throughput release assay development. Ultimately the optimized methods will enable MSCs to become a real therapy and replace more costly and less reliable tests currently in the market.

AUTHOR INDEX

AUTHOR INDEX

AIChE
120 Wall Street, FL 23
New York, NY 10005-4020

ISBN 978-1-5108-6374-3